Spatial Modeling in Natural Sciences and Engineering

Springer

Berlin
Heidelberg
New York
Hong Kong
London
Milan
Paris
Tokyo

Jürgen Friedrich

Spatial Modeling in Natural Sciences and Engineering

Software Development and Implementation

With 101 Figures and 86 Tables

 Springer

Dr. Jürgen Friedrich
Information Technology and Electronics Research Institute
TUBITAK BILTEN
06531 ODTU - Ankara
Turkey
E-mail: *jafriedrich@yahoo.de*

ISBN 3-540-20877-1 Springer-Verlag Berlin Heidelberg New York

Library of Congress Cataloging-in-Publication Data Applied For

A catalog record for this book is available from the Library of Congress.
Bibliographic information published by Die Deutsche Bibliothek
Die Deutsche Bibliothek lists this publication in die Deutsche Nationalbibliographie; detailed
bibliographic data is available in the Internet at <http://dnb.ddb.de>.

Springer-Verlag Berlin Heidelberg New York
Springer-Verlag is a part of Springer Science+Business Media

springeronline.com

© Springer-Verlag Berlin Heidelberg 2004
Printed in Germany

Cover Design: Erich Kirchner, Heidelberg
Typesetting and Layout: Camera-ready by the author

Printed on acid free paper   30/3141/LT – 5 4 3 2 1 0

PREFACE

Why this book?

This text is a **handbook** for students and engineers. It introduces them to the creation and implementation of space-related models, which are models based on geometrical points that are modeled by a *MyPoint* class. Based on a variety of examples formulated during my ten years of teaching and research, the book takes a learning-by-doing and problem-oriented approach to teach the skills needed to build spatial models and then to write a computer program for the model. These procedural skills are rarely taught at universities and many engineers struggle to transfer a model into a computer program. The purpose of this book is to fill this gap. It moves from simple to more complex applications covering various important topics in the sequence: dynamic matrix processing, 2D and 3D graphics, databases, Java applets and parallel computing.

This book is designed to **combine theory and practice** of building models with objects and patterns for spatial applications in natural sciences from a practical point of view by **integrating** object orientation and patterns into space-related model building. All example programs in this book except the Java applets use a **file as the main building block** (*MyFile* as the base class) to create applications containing a parent/multiple child object system as the underlying design structure. Here a file can be of the following type:

- Data text (ASCII) files,
- Graphics or image files,
- Database files, or
- Executable files with "exe" as file extension.

It will be demonstrated in part II of this book how these different file types can be processed by building on **just one base class (*MyFile*)** and retain the general design and implementation of a variety of window programs. The main reason for this is the seamless application of design patterns (mainly the builder, factory method and prototype patterns). For instance, the first example program can simultaneously open and read data text and graphics files. By extending this approach to database and executable files, a simple and powerful design method is at hand to build other wanted applications.

All example programs in part II of this book are **complete extendable medium-sized** applications that address real needs in natural sciences. These examples are more than just scattered pieces of computer code without relevance to the reader who is often confused by the huge amount of material (e.g. most of the Deitel computer books have at least 1000 pages, http://www.deitel.com) and dissatisfied by tiny example programs. Instead, this book tackles a serious problem for many readers today: **how to find relevant information** and parts for buildings models and how to assemble them? This text provides a **solution** by offering a wisely selected and versatile collection of important applications, and by guiding readers through this material using their sequence diagrams. Sequence diagrams are similar to flow charts and show the flow of control and message passing between objects of an application in a timely sequence. They are very useful for explaining in a simple way in which a more complex application works. These explanations in part II of this book are based on "walking through" the steps of a sequence diagram and describing all object interactions and their operations, which are similar to functions and subroutines in Fortran. This also helps a reader to keep the overview and control over an application.

When thinking about the need for a simple, clean and functional method for average model builders, it has become obvious that the present standard for object-oriented system development, the **unified modeling language** (UML), is too complex and confusing for many people who are overwhelmed by the huge amount of UML concepts and diagrams. They do not know how to find what is important for them, and how to build their own models. This is one reason why the UML predecessors were included in this book. They are easier to understand and apply. Other reasons are to help to comprehend the UML because it is a merger of older modeling techniques, and to help to select from the UML what an average model builder just needs in terms of a simple, clean and functional approach (about 10% of the total UML). Simply speaking, two parts are required. Firstly, you have to select an overall architecture for your model and secondly, you need to have a working plan to guide and tell you how to build this model with its architecture. Both parts are provided in this text. The working plan consists of sequence diagrams mentioned before. The overall architecture is a **three-tier architecture** with a graphical user interface as top layer, a processor as middle layer and a data manager as bottom layer.

In this way, this book wants to assist students, engineers and other professionals, who have either no or only little knowledge about OO techniques and patterns, in their desire to create their own and/or alter existing models and computer codes. In doing so, any writer of such a text and any model builder face the same basic conflict: **complexity versus simplicity**. A high-quality model is a compromise of both extremes and its creation requires a lot of skill and experience. In the author's opinion, **learning-by-doing** is the most effective method of knowledge acquisition for building successful models. Therefore, it is strongly recommended to start practicing as soon as possible. Further, this book tries to help the reader by using a **step-by-step** writing approach moving from simple to more complex models and examples. And where suitable, the **history** of covered material is described so that the reader can better comprehend the background and reasons why things happened as they evolved in time. This helps to build a deeper understanding.

Generally, creating and verifying models is a fundamental part of any scientific discipline that comprehends life and uses models in order to reach pre-defined goals. **Models** are representations of something real that can be explored and manipulated to get a deeper insight into its nature. Models can be transferred into products or services that satisfy human needs. Before applying a model, its correctness needs to be tested by sufficient experiments and found to be true within an acceptable fault tolerance (Popper 2002). To make the task easier and less cost intensive, modeling concepts and tools should be appropriate and efficient to do the job. They should be allowed to represent reality as good as possible and narrow the gap between the model and its real counterpart. Object-oriented (OO) methods and patterns belong to the most promising techniques for establishing a more ideal building approach for models.

Finally, this preface has to conclude with a remark about a general discussion regarding model building. Is this activity more an **art**, or does it belong more to the natural sciences with a pure interest in functionality? As in real life, there is no clear answer to this question. Most likely it is a good mixture of both. As a successful car model or a favorite Web page combines functionality with beauty, a high-quality model can also look and feel good!

This book is divided into two parts. Part I, **Object-Oriented Methodology**, consists of three chapters. Chapter 1 introduces the topic with its historical background and key concepts followed by a breadth-first example about OO methodology. Chapter 2 presents different OO modeling techniques including the "unified modeling language" (UML) and patterns. Chapter 3 presents model building with classes, objects, connections, attributes and operations.

Part II, **Model Building with Objects and Patterns**, is made up of seven chapters and applies the contents of part I to spatial problems in natural sciences. All these chapters are **organized in the same pattern**: after an introduction to the fundamentals the model to be created is described by the underlying requirements followed by an OO analysis and design. Then the model's implementation, applied patterns and testing are explained. In more detail, part II introduces the following applications. Chapter 4 deals with a dynamic matrix processor in Visual Basic, chapter 5 with a 2D dynamic data plotter in Visual Basic, and chapter 6 with 3D visualization and animation in Visual Basic. Chapter 7 is about updating legacy programs written in e.g. C/C++, Fortran or Pascal by integrating them into an interactive window shell, thus retaining the investments in these programs. Chapter 8 introduces the reader to a database application for digitizing digital images. Chapter 9 describes another database application: a city map Java applet with a road finder. At the end, chapter 10 introduces parallel computing with Java threads (also called "multi-threading") to determine orbits of earth satellites.

CONTENTS

PART II: Building Models with Objects and Patterns

ACKNOWLEDGEMENTS

First of all, I would like to thank my best friend Jesus Christ for helping me to write this book. I experienced again his words "I am the bread of life. He who comes to me will never go hungry, and he who believes in me will never be thirsty" (John's Gospel 6,35).

Further, I would like to express my thanks to all my colleagues and students who participated in discussions or projects related to topics covered in this book which helped a lot to improve the quality of the contents.

I am very grateful to all those who contributed somehow to the compilation of the book by, for example, information I received through the Internet or by verbal communication. Special thanks go to those who reviewed the book or selected parts of it.

Naturally, my wife and children played a very significant role during the writing time as they encouraged me very often through their love, joy and patience.

Finally, I would like to thank the publisher for taking this book to a wider readership.

All comments or requests can be directed to:

Dr. Jürgen Friedrich
Information Technology and Electronics Research Institute
TUBITAK - BILTEN
06531 ODTU - ANKARA
TURKEY

Tel: +90-(0)312-210 1310 ext. 1169
Fax: +90-(0)312-210 1315
Email: jafriedrich@yahoo.de

CHAPTER 1: Introduction

1.1 Why object-oriented modeling and implementation?

In an ever changing world not only computer hardware and software are experiencing an ongoing process of innovation with more and less expensive functionality, but at the same time every human activity that uses computers is part of this "race for perfection or the best", and faces the same challenge to constantly improve its efficiency and versatility. Modeling and the implementation of models are no exception to this trend. Both are fundamental to any discipline in natural sciences. Modeling incorporates every activity to create a model, which is a representation of a real counterpart that can be explored and manipulated to get a deeper insight into its nature. Gained knowledge can be further applied to reach predefined goals by e.g. creating products or services that satisfy human needs.

Examining how scientific models have evolved in time, one realizes that the first main method being used has been **functional decomposition** in conjunction with reassembling. Hereby, a problem is dismantled into its indivisible functional units for which basic behavioral rules or laws are already known, or if not, needs to be searched for. After applying these rules or laws to each unit, the final model is completed by reassembling each unit according to the conditions found before the dismantling of the considered problem. If this method of functional decomposition together with reassembling cannot be used in practice, it is at least theoretically possible, as it is done in differential and integration calculus. For example, the behavior of a multi-body system is modeled by applying the equations of motions to each body, thus using the concept of functional decomposition and differential calculus. Then the behavior of the whole system is obtained by reassembling it, that is integration over all bodies and time on the basis of their boundary conditions.

Naturally, when the age of digital computers began in the mid 1950's, the **method of functional decomposition** determined very much how the first computers were programmed. The main activity of this method concentrates on mapping the problem to a model based on functions, sub-functions, and functional interfaces. Besides, many other methods and combinations of them have evolved until now. They can be categorized into three basic groups: functional, data-driven, and object-oriented. Their history is briefly summarized in the following table.

Table 1.1. Different methods for modeling and implementation

Beginning	Name	Emphasis on...	Main Concept
Mid 1950's	Functional	Functionality	Functional decomposition
Late 1970's	Data-driven	Data	Data representation
Late 1980's	Object-oriented	Behavior	Object orientation

The evolution of methods for modeling and implementation started with the functional approach in the mid 1950's using functional decomposition to create models and code mainly for processing purposes, which was the most important aspect of computer usage until the 1970's. With growing storage capabilities of computers in the late 1970's, the **data-driven method** emerged. Its main viewpoint is the flow of data and its representation by computers to find the optimum solution to questions like "How can data be most effectively stored, processed, searched, sorted, updated, etc.?" Slowly both approaches (functional and data-driven) developed into what is today commonly called **structured methodology** for modeling and implementation beginning in the early 1980's. Here **methodology** is defined as a package of methods to perform a certain problem-solving strategy, and a **strategy** consists of planned activities to reach pre-defined goals (Norman 1996). As computers grew in power, more complex problems came on the agenda, such as an interactive graphical user interface (GUI), or distributed client/server applications. Soon it was recognized that the functional and data-driven approaches are not very suitable to tackle such problems. Therefore in the late 1980's, both methodologies were basically combined to form the **object-oriented method** by creating objects that are able to encapsulate data, functionality, and behavior. After adding other concepts to this synthesis to build the main concept called "object orientation", a more powerful methodology was born that provides more flexibility and versatility than any other approach so far.

Looking from an outside perspective, one might say that the pendulum of science history is now swinging back. In the beginning of the computer age, computer methodology was heavily borrowing from other scientific disciplines, like in the case of functional decomposition. Nowadays, ideas and techniques from computer science are moving back to other scientific disciplines to bring forth innovation. OO methodology (OOM) is one of them. It allows modeling a problem and implementing its solution more directly in a more complete "three-dimensional picture". The question in the headline of this paragraph **"Why object-oriented modeling and implementation?"** can be briefly answered as follows where the sequence of arguments does not indicate their importance.

- OOM is using natural real-world concepts of organization (e.g. encapsulation, inheritance, polymorphism) for modeling a target problem (or system) and not something gained from a lower dimension, for example, just the processing taking place.

- OOM is a more suitable form of modularization with more self-contained and flexible units generated from a more general system concept.
- OOM offers a more natural system representation and is thus more advantageous for both client/users and designers because they allow simulating the real world more closely.
- OOM allows faster and thus more cost effective adoption to future requirements from client/users because a general concept about a system does not change as fast as more specific characteristics, allowing the reuse of components at a higher level. Therefore, the OOM focus is set more on evolutionary modeling and implementation.
- There is almost no gap between a developed object-oriented model and its implementation in an object-oriented computer program.
- OOM addresses front-end conceptual design issues, rather than back-end implementation issues. Thus, a designer can think in terms of the application domain and concentrate on the implementation until the final stages. Focusing on programming details too early restricts design choices, resulting often in less quality and more design flaws, which are more costly to fix during the implementation.
- Higher degree of modularity and flexibility, and
- Greater clarity, readability, maintainability, and reusability.

Before starting a model and its implementation, one has to decide which is the best methodology to choose for solving a considered problem. This is a very difficult decision with many factors influencing the balance of arguments. Generally, an OO methodology will pay off for complex non-linear problems, where "complex" means a problem containing many different dynamical components with variable behavior and relations. Two examples were already mentioned before: interactive graphical user interfaces and distributed client/server applications. But as long as more linear tasks with a pattern like "data input => computations => data output" need to be solved, it is better to stick to a functional or data-driven approach. It is also good to keep in mind that all methodologies (functional, data-driven, and object-oriented) overlap each other to a very high degree so that anything done with one approach can be used with another one. And there is nothing like a superior methodology. They are just different and well suited for the work they were designed for. A combination of them is also possible, but this very often lacks clarity and consistency. In the end, the final choice depends very much on the type of problem for which a solution is wanted.

1.2 Background of object-oriented methodology

Object-oriented methodology has emerged from various scientific disciplines like group theory, graph theory and combinatory, artificial intelligence including expert systems and neural networks, which try to describe how a more complex system operates. In its beginnings in the 1980's, the OO approach has been mainly used for databases, expert systems, and geographical information systems (GIS). But in the last few years' object-oriented models have also been successfully applied to engineering problems like in robotics (Visinsky et al. 1994) or for the construction of machine parts (Fritzon et al. 1994, Kecskemethy & Hiller 1994). OO methodology was implemented for the first time by Kay (1969) and has brought forth since then many programming environments, among which the most famous are Smalltalk (Goldberg & Robson 1984), C++ (Stroustrup 1987) and Java (http://java.sun.com).

In the future, the OO approach could help us to tackle more challenging problems and model more complex systems (Coad & Yourdon 1991). A **system** consists of a number of interrelated components that work together for a common purpose. There are two types of systems: natural and fabricated (Norman 1996). Natural systems include the human body, the solar system, and the earth's climate system. Fabricated systems are created by humans to reach certain goals or serve pre-defined purposes, as automobiles, airplanes, radios, etc. are doing. In an OO perspective, a system can be defined as a collection of objects mutually interacting in the exploitation of resources and responsibilities (Baujard et al. 1994). The introduced term "**object**" can represent almost everything in our world and we can have different classes of objects. Some examples of what an object can be are material bodies, figures, physical phenomena, people, words, equations, human actions e.g. pressing mouse buttons, etc. Objects can be concrete, such as a pixel on a computer screen, or conceptual, like a priority list in a multi-user system. Generally, objects within an OO methodology are able to perceive and represent the environment in which they are placed: they may communicate with other objects and possess an autonomous behavior, depending on their resources, observations and interactions with other objects. In summary, objects know by themselves who they are, what to do, and how to behave. The OO methodology is very useful to, for example, model the behavior of complex non-linear systems, which addresses many different issues like the:

- Identification of all relevant objects,
- Description, decomposition, and allocation of their tasks,
- Interaction between objects,
- Type of communication and protocol between objects,
- Way to achieve collective behavior in order to perform organized activities and reach a common goal, or
- Choice of an appropriate implementation language and environment.

As already mentioned, one main advantage of an OO methodology is its ability to describe effectively more complex systems and to transfer the model to an object-oriented computer code without major changes, so that the gap between the designed model and the computer implementation is very small during the whole modeling process, allowing very fast design and modifications to achieve convergence. This capability allows an increasing automation, that is a simplification and acceleration of engineering model development, which is otherwise done over and over again for slightly changed or extended problems involving a high portion of repetitive work to solve low-level problems (Fritzon et al. 1994). There seems to be a growing demand among engineers for a higher level modeling- and programming environment to realize something like "tell your computer the problem you want to solve, press a key, go home and wait for the machine to get the answer" (Hardwick & Spooner 1989; Sanal 1994). Object-oriented techniques seem to be more advanced and suitable tools for client/users and designers to come closer to such a dream.

1.3 Key concepts of object-oriented methodology

Object-oriented methodology seeks to mimic the way we form models of the world. To cope with complexities of life, we have evolved a wonderful capacity to generalize, classify and generate abstractions. Almost every noun in our vocabulary represents a class of objects sharing some set of attributes or behavioral traits. OO methodology exploits this natural tendency we have to classify and abstract things. Object-oriented methodology is implemented by object-oriented programming. Supporting and enforcing the use of the following key concepts whereby in an OO terminology, "data" generally corresponds to "attributes", and "functions" are called "methods" or "operations", can characterize both of them. To help the reader to adapt, these OO terms are already used in the following list of **key object-oriented concepts**.

- **Encapsulation** is the combining of attributes with the operations dedicated to manipulate the attributes. Encapsulation allows the hiding of information and is achieved by means of a new structuring and data-type mechanism, which is named "class".
- **Inheritance** is a concept for expressing similarity by which a newly built, derived class inherits the attributes and operations from one or more ancestor classes, while possibly redefining or adding new attributes and operations. This creates a hierarchy of classes like an inheritance tree.
- **Polymorphism** is a concept for changing the behavior of operations when used by different classes. Or in other words, a operation with a given name is shared up and down the class hierarchy, with each class in the hierarchy implementing the operation in a way appropriate to itself.
- **Dynamic** (or late) **binding** is the concept to implement polymorphism by which the physical behavior of an object is prolonged to the adequate moment at run-time of the program.

Encapsulation greatly facilitates the usage of complex data structures to model a complex system, thus allowing for the collection of information of different types that belong naturally together in one class as a single unit, e.g. data of a finite element node containing its number, coordinates, type and value of boundary condition (Forde et al. 1990). Therefore, encapsulation increases the readability and maintainability of the model and code (Sanal 1994). Further, encapsulation sets a boundary between a class and the rest of a program, hiding and protecting its contents from unwanted side effects and thereby making classes more robust. This characteristic of encapsulation is commonly called "information hiding", allowing increasing the modularity of model and code (Norman 1996).

It is also very beneficial to use existing classes when defining new classes. This can be achieved by composition, or by extension of already defined classes using the concept of inheritance. The main purpose of inheritance is to allow sharing of attributes and operations among classes according to a hierarchical relationship, allowing simplifying model and code. From one class, any desired number of class objects can be initialized and used, a process similar to the casting of pieces from a single mould. Each of these copies is autonomous, doing its job on its own, but if it needs some data or code from another class object, the concept of inheritance allows them to share it. Inheritance is of crucial importance for the realization of tree-structured dependencies between class objects. The ability to transfer common attributes and operations of several classes to a base class and inherit them can greatly reduce complexity and repetition within a model. This is one major benefit of the object-oriented approach.

Table 1.2. List of object-oriented terms

Name	Definition / Meaning
Object	An instance of a particular class
Class	A template for objects to encapsulate data, functionality, and behavior
Member	A variable (attribute) or function (operation, service, procedure) within an object
Encapsulation	The binding of attributes and operations into a class of objects
Inheritance	Means that derived classes inherit the attributes and operations from their ancestor classes, while possibly redefining or adding new attributes and operations. This creates a hierarchy of ancestor classes.
Polymorphism	The ability of classes in a hierarchy to share an operation with the same name behaving appropriately to the particular class calling the operation
Late binding	The binding of virtual functions to an object when it is created during run-time
Constructor / Destructor	Operations to create / eliminate objects

Sometimes, a set of class objects may share an operation, but it can only be defined at a specific program level or stage. Thus, this operation can be initialized as a virtual function, but later defined at run-time when it is required. This process is called dynamic (or late) binding. It is part of the polymorphism concept, which makes it possible to adopt an initialized operation to the needs of an object class in the same inheritance tree by adding new features or overwriting existing ones.

Polymorphism simply means that an operation with the same name is able to behave differently in different classes. All relevant object-oriented terms are summarized with their meaning in Table 1.2 (McMonnies & McSporran 1995).

All four concepts are the foundation of any object-oriented methodology, which is a new way of looking at, analyzing, modeling, designing, and implementing models, concentrating on the underlying real-world concepts and principles of a problem or system, rather than processing or implementation details. Thus, there is a general trend in, for example, software development away from programming language issues towards the fundamentals and rules within the application domain, which includes the identification and organization of system components and concepts, their behavior, hierarchy, and relations. This is of great importance for increasing the productivity of modeling and coding as well as reducing the software mountain because general concepts enjoy a longer life span than some implementation details.

In conclusion, object-oriented methodology is a conceptual approach independent from a programming language until the last stages (Rumbaugh et al. 1991). It is a fundamental new way of thinking about and modeling a problem or system, but not a programming technique. Its greatest advantages originate from its capability to create a better and more complete picture of a problem domain. It serves as a more efficient platform for all parts of the modeling and implementation process including concepts, specifications, analysis, design, implementation, testing, documentation, distribution and maintenance.

1.4 Breadth-first approach to OO methodology

The breadth-first approach adopted in this chapter for beginning to teach object-oriented methodology gives exposure to just the essential concepts and elements of OO methodology, but relegates depth of knowledge to later chapters in this book (Nagin & Impagliazzo 1995). Thus, the reader should be able to:
- Understand how an OO methodology works in principle,
- Realize the simplicity and elegance of an OO methodology, and
- Realize the potential and advantages of an OO methodology.

An interesting and scientifically relevant example was chosen to take the reader through all stages of an OO modeling and implementation, starting with the first idea and ending with the final program (or parts of it). The example deals with a second boundary value problem of an accelerated projectile to demonstrate, for example, the benefits of computational experimentation. Naturally, the chosen example cannot be too long and complicated, ensuring an easy-to-follow writing style, which is a major goal of this book.

1.4.1 Basic idea and specifications of an OO example

The basic idea and specifications of the example are the following ones. A basket-ball game between two teams, the "A-City Lions" against the "B-City Tigers", should be simulated by a computer, whereby the *main (or parent) window* of the screen is equally divided into two parts called *child windows* corresponding to two halves of the field with the baskets of each team at both ends. Alternately, players should be positioned by moving the mouse and clicking its buttons, and in a similar way, the speed and direction of a shot should be selected. Then the ball orbit should be immediately drawn, the results displayed and automatically counted. To make it simpler, the problem is reduced to two dimensions in a side view perspective, and certain elements are kept constant like the position and size of the basketball and the field. In summary, the task is to create an interactive window program that takes advantage of its built-in multi-tasking and point-and-click facilities to simulate a basketball game and illustrate the second boundary value problem of an accelerated projectile.

1.4.2 Starting the analysis: what concepts and/or laws are governing the considered problem?

Historically and still today, the accelerated motion of a projectile (here a basketball), beginning at a given starting position (here a basketball player) and ending at a fixed target position (here a basket), is a research topic not only in sports (a little joke), but also in astronomy, geodesy, or ballistics (Falk & Ruppel 1983). Further, it is one of the crucial concepts to grasp in **classical mechanics** and therefore, student instruction should pay adequate attention to it. Mathematically, it is a second boundary value (inverse) problem to Newton's second law of motion, which is an ordinary differential equation of second order in time. If this law is reduced to a two-body problem with only central symmetrical gravity fields, an analytical solution can be obtained in form of Kepler's laws (Heitz 1980). Generally, the second boundary value problem can be solved by using the first one where all starting values are given in one location at the same time in combination with a predictor-corrector technique. This is exactly what basketball players are doing: they perform experiments based on a trial-and-error method by changing their starting values (player position, speed and direction of the ball when shot) in such a way that the ball flies into the basket. For simplicity, only the two-body problem together with a numerical integration scheme is employed, but to make the game more interesting and increase its learning effect, a team should be able to change the position and value of the central gravity mass in its own half of the field within reasonable boundaries after the other team has scored. Thus, the simulated game should use two different gravity fields in one geographical location, including an element of fiction in order to visualize the effects of changing physical parameters and experience alternative realities. This allows the user to experience the consequences of breaking physical laws, providing opportunities for comparative testing of different models and comprehension of their underlying logic (Hennessy et al.

1995). Such a program could also serve as a **computer-assisted learning** (CAL) tool, which can be of great benefit in science education where many students have learning difficulties in acquiring knowledge about fundamental mechanical concepts like gravity, mass or acceleration (McDermott 1990, 1991). Moreover, the OO methodology applied here allows an easy incorporation of other forces such as friction or inertia forces.

1.4.3 OO modeling of a basketball game simulation

The method of creating the object-oriented model for simulating the specified basketball game is the object modeling technique (OMT) proposed by Rumbaugh et al. (1991) which allows an object-oriented analysis of complex systems and builds naturally on the key OO concepts introduced before. The OMT was selected for this breadth-first approach because it is the simplest OO approach to learn in the author's experience, partly due to strong similarities to the structured approach (Sully 1993). But instead of just modeling the data flow as done in the structured method, the OMT uses different projections of varying complexity for the same system like in modern computer drawing programs in order to model the system's information, behavior, and functionality. All projections named **models** depend on each other (McMonnies & McSporran 1995): an object model, which represents the static framework of the system (i.e. what the components are and how they relate to each other) expressed by class and object diagrams; a dynamic model, which describes the events, states, and conditions influencing the system illustrated by state diagrams; and a functional model, which represents the data transformations of the system illustrated by data flow diagrams (where data comes from, what it is transformed into, and where it goes to). As technical drawings are using three different viewpoints (side, front, and top) of the same item, the three models of the OMT are orthogonal parts of one description to represent the whole system. A nice feature of OMT is that the same diagrams and notation can be used for all modeling and implementation stages.

1.4.3.1 Definition of the system domain and boundaries

Before developing the object-oriented model a brief look at the system's functional model gives an idea about the system domain and boundaries, allowing their appropriate definition. Starting with the physical components of the system, a first simple functional model of human interaction with computers is shown in Fig. 1.1.

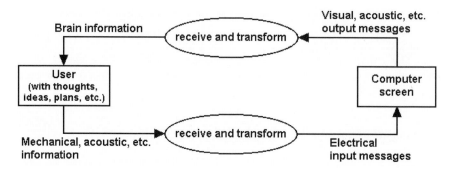

Fig. 1.1. Counter-clockwise data flow diagram for the interaction between user and computer

A user sends mechanical, acoustic or other information based on thoughts, ideas, plans, etc. to a computer which receives and transforms them into electrical input messages. After reacting accordingly the transmittal is reversed; the user receives visual, acoustic, etc. output messages from the computer and transforms them, by using human senses, into brain information. Concentrating only on mechanical and visual means relevant to the application considered here, Fig. 1.1 can be better specified as follows.

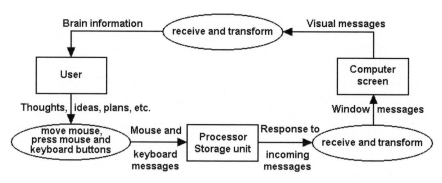

Fig. 1.2. More detailed data flow diagram for the interaction between user and computer

To keep things as clear as possible, the system is reduced to what a user is physically doing: watching a computer screen and using mouse / keyboard as input devices so that Fig. 1.2 simplifies to Fig. 1.3.

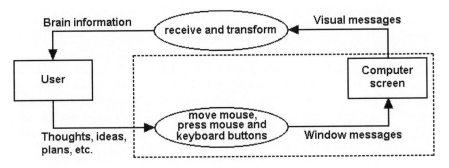

Fig. 1.3. Counter-clockwise data flow diagram with the system domain and boundary as dotted line

The dotted line in Fig. 1.3 marks the boundary of the system domain. Thus, the only directly used physical system components are mouse, keyboard, and computer screen.

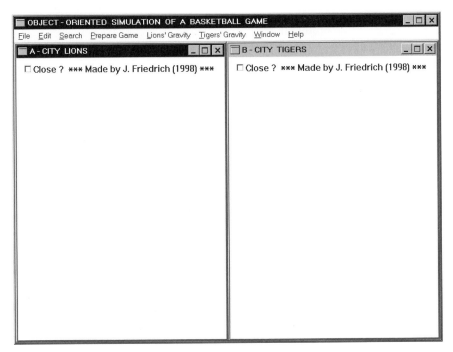

Fig. 1.4. Window display of the OO simulation of a basketball game

1.4.3.2 Object model

The object model identifies and describes the system components and their structure. The system domain is the display of a window program with a main (or parent) window and, for multi-tasking, several child windows (here two are needed), each containing basic window elements like a title bar and control boxes (Fig. 1.4). Moreover, the main window in Fig. 1.4 contains a menu bar for standard window (File, Edit, Search, Window, Help) and application-specific functions (Prepare Game, Lions' Gravity, Tigers' Gravity). Such a screen can be created using window libraries like those provided with C++ compilers for the Microsoft Windows^{TM} operating system, for example, the Borland C++ compiler (1994) used for this program. Thus, classes of this compiler like *TApplication, TMDI-Frame, TMDIClient, TMDIChild* and *TWindow* serve as base classes for self-written (by the programmer) derived classes needed for the application-specific functions completing the class diagram shown in Fig. 1.5 (Class names in the text are written in italics; classes containing *My* in their name are self-written; other classes are taken from the compiler's window class library).

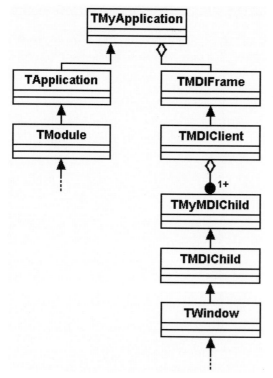

Fig. 1.5. Class diagram of the OO simulation of a basketball game

Fig. 1.5 displays only a very basic object model with the most important classes. The complete model of the final program is much more detailed and can be viewed by using e.g. the class browser of the used compiler. It includes, for example, classes to create and handle file menus and dialog boxes.

TMyApplication is derived from *TApplication* and *TModule* (indicated by an arrow for a one-to-one relationship), both responsible for the basic behavior of a window program including their initialization and message processing. A constructor of TMDIFrame sets up the main window or frame of TMyApplication, which is a part of *TMyApplication* (indicated by the diamond symbol). TMDIClient manages the child windows inside of the frame. All *MDI* classes belong to the multiple-document interface (MDI) that takes care of e.g. constructing and handling parent and child windows. *TMDIClient* creates child windows by calling a *TMyMDIChild* constructor. Thus, a one-to-many relationship exists between *TMDIClient* and *TMyMDIChild* (indicated by the diamond and circle symbol), a class derived from *TMDIChild* originating from *TWindow*. The *TWindow* class provides all the window-specific functionality for a child window. *TWindow* and *TModule* build upon other window library classes not mentioned here (for further information see Borland (1994)). In this example *TMyMDIChild* is of special importance because it supplies everything needed for the target model and code.

1.4.3.3 Dynamic model

The system domain in Fig. 1.4 consists of a main window including menu bar items and two child windows as system elements. The feature of a window program which makes it so effective is its ability to change the state (activated or deactivated) of any system element by just positioning the mouse on a considered element and pressing a mouse button (normally the left one), which is the event causing the element's state to change. This is automatically performed by virtual event message response functions of the built-in window library classes (Borland 1994).

The execution of menu bar items or one of their corresponding pop-up choices takes place in a similar way. If the item is provided by the compiler's window class library like for File, Edit, Search, Window, or Help, the response happens accordingly to the built-in library functions. The Help item is placed into this category because only its contents need to be adjusted to the present application. A self-created item such as Prepare Game, Lions' Gravity, or Tigers' Gravity is linked to a virtual function of *TMyMDIChild* through one command message (CM) identifier given to both the menu bar item and the corresponding member function when they are declared. Virtual window message (WM) response functions and identifiers process analogously mouse moves and presses inside of an active child window. Both CM and WM functions are essential tools of the designed dynamic model shown in the state diagram in Fig. 1.6.

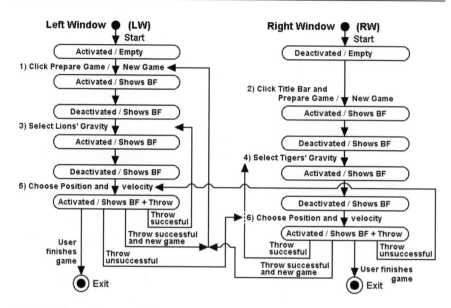

Fig. 1.6. State diagram of the OO simulation of a basketball game

At the beginning of the game after starting the program, the left window (LW) in Fig. 1.4 entitled "A-CITY LIONS" is activated shown by its highlighted title bar, and the right window (RW) for the other team "B-CITY TIGERS" is deactivated. Both windows are empty. The game begins after the menu bar item Prepare Game / New Game is clicked in both windows, which sets the score counters to zero and displays the basketball field (BF) (step 1 and 2). Step 3 and 4 allow both teams to choose their gravity location by selecting the menu bar items Lions' Gravity and Tigers' Gravity. Then either team can start the game, for example, the Tigers. They move the mouse to the Lions' LW, press the left mouse button down (only this button is used throughout the whole program) at the wanted position of their player and keep the button down until the mouse is moved to and released at a second position (step 5). The difference between the first and second position $P_i^1 - P_i^2$, $i \in \{x, y\}$, defines the ball's velocity vector used to compute its orbit, immediately shown in the LW after releasing the mouse button. Then the Lions start their game by moving the mouse to the Tigers' RW, following the actions of the 5th step in an analogous way (step 6). Steps 5 and 6 are repeated alternately until one team scores, so that the score counter at the top of each window is incremented by two (Fig. 1.8). In order to make the next attempt harder, the other team is allowed to change its gravity location and value (step 3 or 4). The game is finished after reaching a score or time limit. Going back to the first step can start a new game.

1.4.3.4 Functional model

The functional model completes the object and dynamic model by concentrating on the flow of data between stores and processes. Processes are all operations that transform data inputs into data outputs. Employing the sequential steps of Fig. 1.6, the simplified data flow diagram in Fig. 1.3 can be specified as shown in Fig. 1.7. The data is only generated by the ideas, plans, and decisions of the users (the two teams Lions and Tigers) wanting to perform the visual interactive simulation of a basketball game by adequate processes (mouse moves and mouse/keyboard button presses) according to the object model explained and shown before in Fig. 1.6.

Therefore, all data comes from the users (1st data store) and flows to the computer screen (2nd data store) where it is stored as visual data as screen contents. There it lasts and remains visible until it is changed by the window display as response to later commands, for example, a newer one replaces the old graphics of a basketball field.

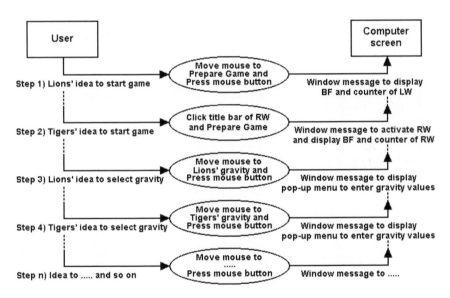

Fig. 1.7. Data flow diagram of the OO simulation of a basketball game

1.4.4 OO implementation of a basketball game simulation

For implementing the created model, a programming language has to be selected, which always includes a compromise because there is nothing like a "perfect" programming language yet. Here C++ was chosen because:

- It is a multi-purpose language capable of solving both high and low-level computer problems.
- It offers together with ANSI C a large selection of existing libraries and other resources (e.g. text books) for many different applications, e.g. for numerics, databases and graphics (e.g. OpenGL).
- It runs on almost every hardware and operating system, and
- Offers OO facilities, such as dynamic link libraries (DLLs) or CASE tools.

The implementation of virtual CM and WM response functions, which represent the largest portion in the final program, are just briefly explained in this paragraph and illustrated by the following C++ code (for further information see e.g. Borland (1994) or Heiny (1994)).

Table 1.3. Example C++ code for virtual CM and WM response functions

```
01    ...                               // Declaration of class TMyMDIChild
02    class TMyMDIChild : public TWindow { public:
03    TmyMDIChild (TWindow &parent,  const char far *title);    // Constructor
04    ~TmyMDIChild ( );                                         // Destructor
05    int x,y, vx,vy;           // Four integer numbers for position and velocity
06    ...                            // Declaration of virtual functions:
07    virtual void EvLButtonDown (UNIT, TPoint&); // Left mouse button down
08    virtual void EvLButtonUp (UNIT, TPoint&);    // Left mouse button up
09    DECLARE_RESPONSE_TABLE (TMyMDIChild);
10    };                        //Declare response table
11    DEFINE_RESPONSE_TABLE1(TMyMDIChild, TWindow)
12    EV_WM_LBUTTONDOWN,      // Definition of response table
13    EV_WM_LBUTTONUP,
14    END_RESPONSE_TABLE;
15    ...                            // Definition of virtual functions
16    void TMyMDIChild::EvLButtonDown (UINT, TPoint& Coord) {
17    HDC hdc = GetDC (HWindow);   // Get window device handle
18    x = Coord.x;  y = Coord.y;     // Store window coordinates of mouse in x, y
19    ReleaseDC (Hwindow, hdc);   // Release window device
20    };
21    void TMyMDIChild::EvLButtonUp (UINT, TPoint& Coord) {
22    HDC hdc = GetDC (HWindow);       // Get window device handle
23    Vx = x - Coord.x;  vy = y – Coord.y;  // Store velocity components in vx, vy
24    ReleaseDC (Hwindow, hdc);       // Release window device
25    };
```

Here, pressing the left mouse button down at a first position stores the mouse coordinates within the activated window in the integer variables x and y. Releasing the same button at a second position computes the ball's velocity components vx and vy as the difference between the first and second mouse position P_i^1 - P_i^2,

$i \in \{x, y\}$. Both virtual functions "EvLButton-Down" and "EvLButtonUp" of *TMyMDIChild* respond to the incoming window messages EV_WM_LBUTTONDOWN and EV_WM_LBUTTONUP defined in the response table of *TMyMDIChild*. "Coord" is an instance of class *TPoint*, another built-in window class of the compiler. It contains the mouse coordinates of the last event when a mouse button was pressed. "UINT" is an identifier for combined keyboard and mouse events like "Shift + Double-Click", but here not used. A typical screen contents after starting a game looks as is shown in Fig. 1.8 (Download latest version at http://www.simtel.net/pub/pd/53592.hmtl).

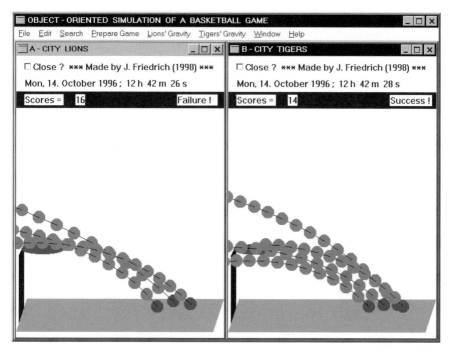

Fig. 1.8. Screen shots after starting the OO simulation of a basketball game

As this OO simulation program of a basketball game illustrates, a window program is organized around events and messages within an object-oriented programming environment. All window interface elements like child windows, scroll bars or dialog boxes are objects that combine data with functionality. For example, pressing the OK button in a dialog box may be connected to a virtual function that verifies the entered data and passes them on to some other variables. Such a style of organization can be modeled by, for example, the **responsibility-driven approach** of Wirfs-Brock et al. (1990). The authors have basically divided all involved objects into clients and servers, such that the servers provide some services laid down in a server-specific contract, which can be invoked by clients, who do not have to know how the services are carried out. The main characteristic of this client/server model is the emphasis on actions within object relations instead of representations. The object-oriented model for the basketball simulation program utilizes this responsibility-driven approach and divides all objects into two categories: **state objects**, which are able to hold information, and **acting objects**, which receive this information from a set of state objects, perform their task, and transfer all their knowledge back to the same or another set of state objects, or, if necessary, to some other type of objects. Thus, state objects play the role of clients, acting objects the role of servers, which can be also regarded as mappings between two sets of state objects. Note that the concept of treating state- and acting objects as separate, autonomous entities is very similar to the understanding of nodes and elements in boundary or finite element theory (Brebbia 1978, Zienkiewicz 1977). But the difference is that the two categories of objects introduced here are made visible and approachable, allowing the designer to use them as building blocks of a **"unit construction system"** for modeling other interactive multi-object applications, thus increasing reusability and maintainability.

Such an object-oriented modeling and implementation style benefits very much from the built-in window libraries. For example, both the C++ "Object Windows Library" (OWL) from Borland or the "Microsoft Foundation Class" (MFC) contain over one thousand functions, thus indicating the great potential of such applications. Two other benefits of the introduced approach are the capability of parallel processing on personal computers (PCs) by using multi-threading (a *thread* is a block of code in a computer program), and the design of faster 32-bit programs for PCs.

Regarding the implementation of object-oriented models, another new programming language called Java ^{TM} by Sun Microsystems Inc. already became so popular and widely used since its introduction in 1995 that the amount of OO models implemented in Java is sharply increasing. This book introduces two Java applets at the end of part II: a city map applet with a road finder in chapter 9 and in chapter 10, an applet for parallel computing with Java threads (also called "multi-threading") to determine orbits of earth satellites.

CHAPTER 2: Overview of Different Object-Oriented Approaches

2.1 Common characteristics of different OO approaches

All different object-oriented modeling approaches have several characteristics in common. They are based on object orientation and the OO key concepts introduced before. All of them are also using a graphical notation with textual support to augment their graphics. Moreover, the final outcome of each method to represent a system's information and behavior should be more or less identical and independent from the chosen approach. This has to be the case in order to guarantee consistency.

What is different is the path from start to finish, or the perspective under which a system is modeled. Depending on the viewpoint of an observer, s/he concentrates on one aspect of the system in the beginning before going over to another part. Somebody else might do it in just the opposite way. One person might view data or information as the most important. But another person could disagree and stress more its behavior, for example. Thus, the chosen approach to create a model is very much a question of personal preference (more artist or engineer?) and not objective at all. One system can be modeled in different ways, but the result should be about the same.

Therefore, there is nothing like a better or superior modeling approach. Generally, they are somewhat different but well suited for the job they were designed for. Another factor of influence regarding the choice of a modeling method is the question whether somebody likes or is used to working with different projections to represent a system, or whether s/he prefers a more complete picture of the same system. For example, technical drawings use three different viewpoints (side, front, and top) as three orthogonal projections of one object. Someone might like this more than a three dimensional view of the same object because such a picture can be too complex and overloaded with information. For another person a three dimensional model could be just right and not too complicated at all.

Concluding, the preferred style to represent a model, either in projections or in a more complete form, is of great importance for choosing an object-oriented mo-

modeling approach. For the projection-type, the method of Booch (1994) and the "object modeling technique" (OMT) by Rumbaugh et al. (1991) are examples. The other type is used in the method of Coad & Yourdon (1991). In this book, the latter method and the OMT will be explained in more detail, whereas Booch's method will only be summarized. The reason is just a practical one because it is enough to study just one projection-type method in depth. Further, Booch's approach and the OMT have strong similarities so that, together with the OOSE/Jacobson method, they were combined to form the "unified modeling language" (UML), the present standard for OO modeling (Booch et al. 1998). The UML will be further described in this text where it contains new parts or differences to Booch's method and the OMT. In this way, the history of different approaches is used again to better understand the rather complicated UML.

As mentioned before, all methods are based on object orientation and its key concepts introduced in chapter 1 together with a breadth-first example. *What does object orientation mean again?* Simply speaking, it is a modeling approach based on a synthesis of functional and data-driven decomposition in conjunction with reassembling, which can be called **object-oriented decomposition** (compare Table 1.1). Thus, the modeling process starts with the search for and identification of objects that encapsulate data and functionality at the same time. Of course, the objects within a problem have to obey or be consistent with the basic behavioral laws ruling the considered problem. After identifying all involved objects, their reassembling begins by establishing the connections and responsibilities of each object.

How is it possible to find all required objects in an easy and fast way? How can the decomposition and reassembling of objects be effectively done in a consistent and straightforward manner? Answers to such questions can be traced back to the human ability to classify, generalize, and draw abstractions. There are **four basic rules** humans are constantly applying from the time they enter elementary school (Norman 1996):

- **Rule I:** *Classify different objects according to their appearance (data), their behavior (functionality), or both.* For example, a house is very different from an automobile regarding both characteristics, whereas the differences between a pocket calculator and a PC are also obvious, but much smaller. Thus, every classification depends on the applied degree of abstraction.
- **Rule II:** *Distinguish between a whole object and its different components.* An automobile, for instance, consists of wheels, engine, chassis, etc., while a house is made of walls, windows, doors, stairs, and so on.
- **Rule III:** *Sort and group different objects into categories or classes.* A class can be defined as an abstract collection of objects. Examples are the sorting of different vehicles (bicycle, motorbike, car, lorry, bus) into a vehicle class, or the grouping of different computers (palmtop, laptop, desktop, workstation, mainframe) into a computer class.
- **Rule IV:** *Describe the relations and interactions between different classes and their objects.* A class of people, for example, is only using laptop computers,

whereas another people group prefers workstations. Almost everything in nature has some type of connection to another class of objects.

Our daily practice and understanding of these rules is a strong argument for object-oriented approaches. The basis of these methods is not completely new and familiar to us. To comprehend and apply these four rules is of crucial importance for being successful in object-oriented model building. But experience has shown that many people have a lot of difficulties to manage and use these rules when more complex and challenging problems need to be solved. Maybe the reason is that they find it hard to visualize and then materialize what they are unconsciously doing. Or they have a tough time changing paradigms when their thinking is already so much geared towards other than object-oriented concepts when it comes to building models. Like so many other things in life, only persistent training seems to be able to change this situation.

2.2 A simple example for illustration purposes

Before starting to describe Booch's method, it is helpful to introduce a small example of an object-oriented model for illustration purposes, which can be easily used with any approach, described by some basic object-oriented notation taken from Coad & Yourdon (1991). The example deals with a simple digital terrain model (DTM) analyzed from an object-oriented perspective while applying rules I-IV introduced before. The starting point of an object-oriented DTM is a terrain point seen as one representative or instance of a much larger class of objects, which can be named *Point*. *Point* is the general class name of all terrain point objects. The concept of a point is not only fundamental to geosciences, but also very essential to any other scientific discipline where geometry related data is used. In order to identify a point uniquely, it is labeled by a number, name, or code referenced as pointID here. Further, some type of coordinates describes a point's location. Apart from other functions the pointID and coordinates need to be read from an input device, written to an output device, and displayed as graphics. Each of these services can be managed by different services called read, write, and display. All the introduced information can be encapsulated in the class *Point* as shown in Fig. 2.1.

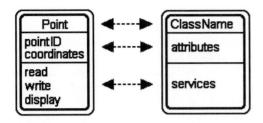

Fig. 2.1. Class *Point* with attributes and services according to the Coad & Your-don (1991) notation

Fig. 2.1 shows the **class-with-objects** symbol introduced by Coad & Yourdon (1991) for the *Point* class with its class name at the top of the capsule, its attrib-utes (pointID, coodinates) in the middle, and its services (read, write, display) at the bottom. Using such a symbol ensures a good compromise between detailed enough information on one side, and simplicity on the other side. Where wanted or necessary, attributes and services can be hidden, as was done in the breadth-first example of chapter 1. A class-with-objects is the most essential part of any object-oriented model. While a class is an abstract collection of objects, an object itself is the actual instance or representative of a class.

Once the class *Point* has been defined, the attributes and services can be reused whenever a new object of the class is created. For example, if a number of n ter-rain points named P-i, i ∈ {1,2,3,..,n}, is considered, each P-i object instantiates all attributes and services defined by class *Point* according to the encapsulation concept so that all of its information is packaged under one name and can be re-used as one component in the overall model. This is illustrated in Fig. 2.2.

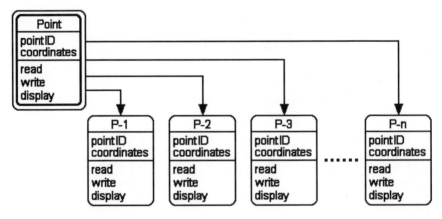

Fig. 2.2. Every P-i object, i ∈ {1,2,3,..., n}, inherits all attributes and services from class *Point* in the Coad & Yourdon (1991) notation

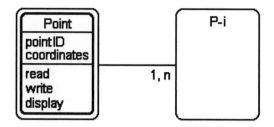

Fig. 2.3. Class-to-object connection between class *Point* and P-i objects, $i \in \{1,2, 3,...,n\}$ in the Coad & Yourdon (1991) notation

Fig. 2.2 can be simplified by using a single line between class *Point* and the P-i object labeled with "1,n" as shown in Fig. 2.3. "1,n" represents the lower and upper bound of the number of objects (range). For simplicity reasons the interior of the P-i object was left empty in Fig. 2.3.

Apart from a certain number of surface point objects the object-oriented DTM requires at least two more classes. A second class named *GraphicSystem* serves to initialize the computer graphic system to enable the machine to pass any type of graphic to screen, printer, or plotter. This class incorporates all necessary attributes and services for this task, depending on the individual hardware and software configuration. It requires, for example, the screen coordinates of the graphic center, and services to draw the coordinate frame. A third and for now last class *Topography* is derived from the other two classes and uses their attributes and services to create a terrain model. This also requires its own attributes, such as the scale of the DTM or the view angle, and services, for example, to select the type of interpolation function and to draw the connection between different terrain points. The described class hierarchy is displayed in Fig. 2.4.

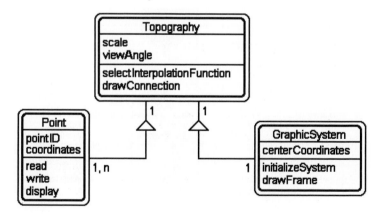

Fig. 2.4. Class hierarchy of a simple digital terrain model in the Coad & Yourdon (1991) notation

Here a **whole-to-part connection** exists between the *Topography* class as the whole on one side, and the two base classes *Point* and *GraphicSystem* as parts of class *Topography* on the other side. Placing the whole higher and the parts lower produces an easier-to-understand model. This also portrays that a whole has normally some number of different parts. The symbol for a whole-to-part connection is a single line with a triangle in between labeled with the number or range of objects (Coad & Yourdon 1991). The triangle always points in direction of the derived whole class, here *Topography*. Concluding, the created DTM shown in Fig. 2.4 can be briefly described as follows. A class *Point* with one to n objects and another class *GraphicSystem* with one object pass their attributes and services on to a class *Topography* with one object that allows to select an interpolation function, and then draws connecting lines between different terrain point objects. Before the object of class *GraphicSystem* initialized the graphic system and displayed a coordinate frame.

2.3 Booch's object-oriented method

Booch (1994) introduced a projection-type object-oriented method based on four different models as illustrated in Fig. 2.5, capturing different interrelated views of the same problem/system under consideration.

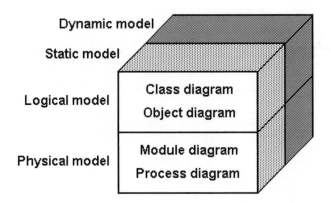

Fig. 2.5. Different models of Booch's method with corresponding diagrams

The logical model describes the structure and mechanisms of the problem/system whereas the physical model contains the concrete software and hardware composition to implement the developed solution. The static and dynamic models capture the static and dynamic behavior of the problem/system. Each model uses different diagrams as shown in Fig. 2.5 and in a sequence as they are listed below.

1. **Class diagrams** describe classes, their hierarchy, and their relations with each other.

2. **Object diagrams** represent created objects as instances of classes. Classes and objects are allocated to specific software components as part of a physical model.
3. **Module diagrams** are used to describe these software components.
4. **Process diagrams** allow specifying how software components are allocated to the available hardware components.

Further, the dynamic model uses state-transition (or shorter just state) diagrams and interaction diagrams (also called scenarios) for modeling the dynamic behavior of a problem/system in the same way as it is done in the "object modeling technique" (OMT) explained in the next paragraph. Because of this and other similarities both methods became major parts of the "unified modeling language" (Booch et al. 1998), the present modeling standard.

In difference to the class-with-objects notation by Coad & Yourdon (1991) used for the last example in the paragraph before (Fig. 2.4), Booch's method separates classes from their objects. In the beginning of the modeling process, the class diagram is created with an emphasis on the selection of classes and their relationships. Booch (1994) prefers to use a cloud symbol to represent a class in this diagram but mentions a rectangle as an alternative. In this book, the rectangular class symbol is chosen for reasons of simplicity as shown in Fig. 2.6.

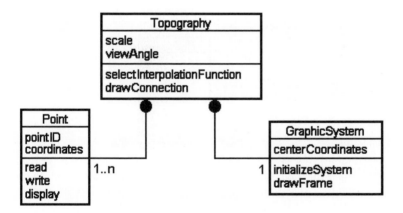

Fig. 2.6. Class diagram of the DTM in the Booch notation

A single line and a filled circle on the side of class Topograghy as the whole illustrate the whole-to-part connection in Fig. 2.6. The labels "1" and "1..n" indicate the number or range of objects as was already introduced in Fig. 2.4. Generally, a class diagram in Booch's method does not explicitly show the attributes and services of a single class, as it is done in Fig. 2.6.

This information is normally transferred to a class template, which is no actual part of the class diagram but is created as a supplement to it. Comparing Fig. 2.6

with Fig. 2.4 reveals that both express, apart from notation differences, the same class model.

Classes were defined as abstract collection of objects within the overall modeling process of a system that exist regardless of its implementation and execution within a computer program. Objects, however, are instances of classes created and deleted dynamically during the run time of a program. Thus, a notation is necessary to capture this dynamic behavior of objects. The object diagram is a tool to do that. It describes the life cycle of objects from birth to death, and incorporates, for example, the services that are invoked, and the messages that are passed. The basic notation of an object diagram is illustrated in Fig. 2.7 where the connecting lines between objects of class *Point, GraphicSystem* and *Topography* are terminated by a labeled square that indicates the data type contained within a message. "F" stands for a field that is shared by both objects. The arrows parallel to the connecting line indicate the type of synchronization between objects, for example, synchronous, asynchronous, or time-out. A crossed-out arrow means a synchronous passing of data and message.

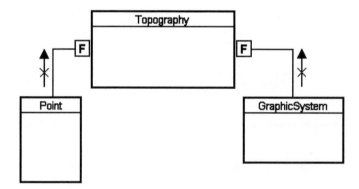

Fig. 2.7. Basic notation of an object diagram as part of the DTM

Similar to the class diagram, the object diagram can be supplemented by an object template containing its definition and the messages that the object is sending and receiving.

A module diagram of Booch's method allows objects to be packaged into modules (files) to be able to reuse them more effectively. The module diagram uses different symbols for main program, specification, body, and subsystem modules. For example, Fig. 2.8 shows a module diagram for the DTM using the body symbol.

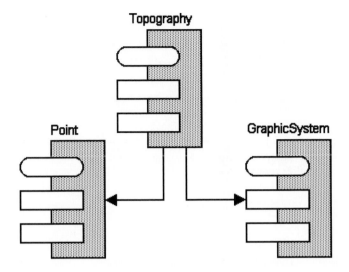

Fig. 2.8. Basic notation of a module diagram as part of the DTM

In Fig. 2.8, the connecting arrows indicate dependencies between different modules. The module at the origin of an arrow depends on the module at the tip of the arrow. The highest-level module contains an object of class *Topography*. When a higher degree of abstraction is used, the whole DTM can be condensed to one module entity called "Point-Engine", for example.

For any designed model implemented on a computer, the resulting code is executed by different hardware components. Object-oriented models are no exception to this. The process diagram allows for a system to specify how software components are allocated to the available hardware components. It is used to depict the processors (shaded boxes), devices (white boxes), and connections (lines) that define the hardware architecture of a computer system. An example is shown in Fig. 2.9.

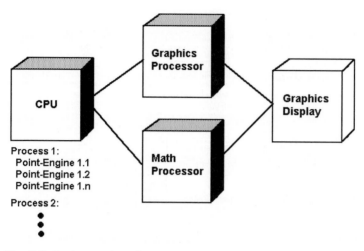

Fig. 2.9. Basic notation of a process diagram

A more detailed introduction to Booch's method as presented so far is beyond the scope of this book. The interested reader is referred to Booch (1994), where a variety of case studies and examples are provided. Concluding, Booch's method was introduced as a projection-type approach that differentiates between classes as the model's static structural parts, objects as its dynamical parts, modules and processes as its functional parts. Another object-oriented method already applied in the breadth-first example of the first paragraph uses the same type of organization. It is described in the following paragraph.

2.4 The object modeling technique (OMT)

The object modeling technique (OMT) proposed by Rumbaugh et al. (1991) allows an object-oriented analysis of complex systems and builds naturally on the key OO concepts introduced before. The OMT uses different projections of varying complexity for the same system like in modern computer drawing programs in order to model the system's information, behavior, and functionality. All projections named *models* depend on each other (McMonnies & McSporran 1995): an **object model**, which represents the static framework of the system (i.e. what the components are and how they relate to each other) expressed by class and object diagrams; a **dynamic model**, which describes the events, states, and conditions influencing the system illustrated by state-transition diagrams; and a **functional model**, which represents the data transformations of the system illustrated by data flow diagrams (where data comes from, what it is transformed into, and where it goes to). As technical drawings are using three different projections (side, front, and top) of the same item, the three models of the OMT are orthogonal parts of one description to represent the whole system. Thus, they are not independent from each other. Each model contains references to entities in other models. For example, operations are integrated in objects as part of the object model but more fully developed in the functional model. Interconnections between the different models are limited and explicit. In the author's experience, the OMT is very easy to understand and learn, partly due to strong similarities to structured methods as can be seen from the following table (Sully 1993).

Table 2.1. Diagrams of structured methods and the OMT

Model area	Structured methods	OMT
Information structure	Entity relationship diagram	Class / Object diagram
Time behavior	State-transition diagram	State-transition diagram
Work or Functionality	Data flow diagram	Data flow diagram

The contents of corresponding diagrams are very close to each other and in fact identical for the area of work or functionality. Naturally, such similarities also exist to other object-oriented techniques. The different diagrams of structured methods can be of great help in the object-oriented modeling process, especially for finding classes and objects (Yourdon & Argila 1996). What follows is a more detailed description of the OMT together with examples taken again from the DTM. Each new concept is introduced by a definition, explanation, notation and an example.

2.4.1 Object model

The object model is the foundation of the OMT and the basis for the actual model to be designed. It describes the static structure of a system or problem under consideration in terms of classes and their relations corresponding to real-world entities. The object model provides the basis on which the dynamic and functional models are placed. It captures how the system or problem is organized and working from an object-oriented viewpoint, whereby the focus is on objects collected into classes, the identity of objects, their relationships to other objects together with their attributes and operations. In the OMT, a class is graphically shown by a rectangle, and services are called "operations". This is illustrated in Fig. 2.10.

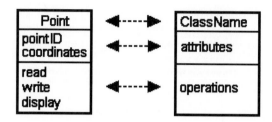

Fig. 2.10. Class *Point* with attributes and operations in the OMT notation

Generalization in the OMT is defined as a generalization-specialization relation between classes, and is applied by connecting two classes with a line together with a triangle, pointing to the general class named "superclass". Specialized classes or "subclasses" are placed at the other end of the line. For example, *Point* could be the superclass of two subclasses *GravityPoint* and *TriangulationPoint*, as shown in the following **class diagram** (Fig. 2.11).

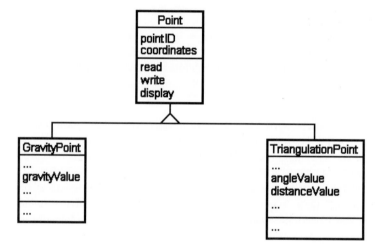

Fig. 2.11. Class diagram of the DTM in the OMT notation

The generalization of classes naturally expresses the key object-oriented concept of **inheritance** explained in paragraph 1.3, that is, attributes and operations of the generalization class are shared by the specialization classes.

Another diagram used in the object model is the **instance diagram**, which shows how objects relate to each other. An instance or object of a class is represented with a rounded rectangle with the class name displayed in parentheses and the attributes listed below. Fig. 2.12 shows an instance diagram for objects of class *Topography* and *Point*.

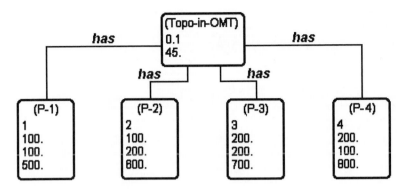

Fig. 2.12. Instance diagram for parts of the DTM

In this instance diagram, one object of class *Topography* called "Topo-in-OMT" with a scale of 0.1 and a view angle of 45 has four points, each an object of class *Point* named P-i, i ∈ {1, 2, 3, 4}, with i represents pointID and three Cartesian coordinates x,y,z.

Links and associations are essential parts of the object model in the OMT because they are the means for establishing relations between classes and objects. A **link** is defined as a connection between objects. In Fig. 2.12, for example, Topo-in-OMT has P-i links an instance of class *Topography* and objects of class *Point*. An **association** is defined as a group or a class of links with common structures. For example, Fig. 2.4 shows an association that class *Point* is part of class *Topography*. Therefore, a link is an instance of an association in the same way that an object is an instance of a class. An association can be bi-directional so that the inverse association is also valid; such as class *Topography* contains class *Point*. The OMT notation for an association is a line between classes in the class diagram, whereas the link notation is a line between objects in the instance diagram. Both association and link names are written in italics above the line, as it was done in Fig. 2.12. An association or link name can be omitted, if the meaning of the connection is obvious, such as in Fig. 2.11.

In class diagrams, it is useful to indicate the number, range or subset of involved objects. **Multiplicity** defines specifiers of objects (symbols, text, or numbers) in class diagrams. It informs about how many objects of one class are linked to how many objects of another class. For instance, the solid ball symbol is placed next to a class symbol at the end of a link to indicate that many objects of this class are involved in the connection under consideration. Fig. 2.13 displays different options for a link between a *Subclass* and a *Superclass* together with the various symbols for multiplicity. All these forms are part of generalization.

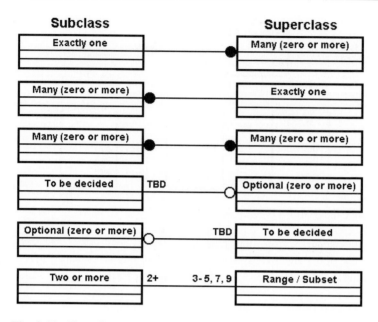

Fig. 2.13. Class diagram for different multiplicity symbols

Apart from the solid ball symbol, the other symbol used in class diagrams of the OMT are the empty ball symbol for an optional number of objects, as can be seen in Fig. 2.13. The other class in this link is marked by the text "TDB" or "To be decided", meaning that an object of this class maybe created or not, depending on the optional object of the other class. The bottom case in Fig. 2.13 uses numerical specifiers to depict the number of involved objects. For example, "2+" represents two or more objects whereas "3-5,7,9" is the range and subset of numbers for the objects to be created. Using numerical specifiers, for instance, Fig. 2.11 could be rewritten in OMT notation in the same way as it is done in Fig. 2.4. The only difference would be that rectangles would replace the class-with-objects symbols of the Coad & Yourdon (1991) notation in Fig. 2.4. Thus, there is no difference in principle between both methods in this aspect.

"One-to-many" and "many-to-one" associations (Fig. 2.13) may have a restricted multiplicity. The OMT tool to include such restrictions is a **qualified association** defined as a class of links with qualifications or conditions. Among all possible objects the qualified association detects those ones that fulfill its conditions. Its symbol is a small box drawn at the end of the connection line next to the class it qualifies. An example for the DTM is given in the following class diagram (Fig. 2.14).

In Fig. 2.14 the qualified association "pointID > 100" restricts the connection between class *Topography* and class *Point* so that only objects of *Point* with a pointID greater than 100 are linked to the object of *Topography*.

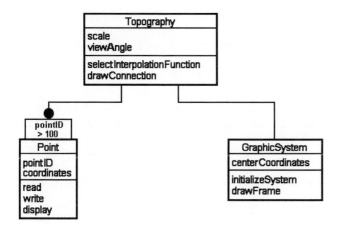

Fig. 2.14. Class diagram of the DTM with qualified association

Aggregation is defined as a "whole-to-part" relationship between a superclass (or assembly class) representing the whole, and only one subclass. A superclass can have many aggregations at the same time. Aggregation is a special type of association and an alternative to other forms of generalization as given in Fig. 2.13. An aggregated superclass is especially useful when its subclasses serve as more independent components and building blocks, which do not change the superclass. But if subclasses and superclass are interdependent, it is recommended to apply "many-to-one" links or other appropriate forms of generalization shown in Fig. 2.13. The notation for an aggregation is the same as for an association, except a diamond symbol is placed at the end of the connection line next to the superclass. An aggregation example is provided for a "geographic information system" (GIS) in Fig. 2.15.

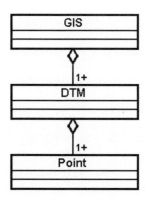

Fig. 2.15. Aggregation used in a GIS model

In Fig. 2.15 class *Point* is part of class *DTM* which is again a part of class *GIS*. Class *Point* and *DTM* have one or more objects. An outstanding property of aggregation is transitivity, meaning that, if *Point* is a part of *DTM* and *DTM* is part of *GIS*, then *Point* is part of *GIS*. Aggregation is also characterized by anti-symmetry, that is, if *Point* is a part of *DTM*, then *DTM* is not part of *Point*.

The object model of the OMT owns other features that may not be necessary for simple models, such as link attributes or abstract and concrete classes (Rumbaugh et al. 1991). A **link attribute** is defined as a property of a link in an association. Its symbol is a text box attached by a loop to the connection line of the link. It may include, for example, access specifications. An **abstract class** is defined as a class without own objects but whose subclasses have direct objects, while a **concrete class** is defined as a class with its own objects. It is possible, for example, that a concrete class can have abstract subclasses as part of a larger application.

2.4.2 Dynamic model

While the object model focuses on the static components of a system, the dynamic model concentrates on its behavior and how it is changing in time or under different circumstances. It describes the dynamic characteristics of a system that cause the system to change: events taking place, a certain sequence of events, states that determine how events can bring forth change, the organization of events and states in conjunction with the dynamics ruling them. The dynamic model captures control over the considered system and how it can be manipulated.

As graphical tools the dynamic model uses state-transition diagrams to represent its contents. State-transition diagrams display the permitted states and events happening in a system. They also allow referring to another model. For example, events in a state-transition diagram correspond to operations in the object model.

An **event** is defined as an action at a certain point in time that causes the state of an object to change. A **state** is defined as the attribute values and links of an object. For instance, clicking the left mouse button over a deactivated window causes the window to change its state from "deactivated" to "activated". The clicking of the left mouse button is the event, whereas the properties "activated" and "deactivated" are the states of the window. In theory, an event has no duration with a physical life span that is negligible compared to the time behavior of the rest of the system. An event is unidirectional, meaning that information is only passed into one direction without returning a result. If a response to an event is needed, another event must take place to do that. If two events are considered and they have no effect on each other, then they are causally unrelated or **concurrent**. The opposite are causally related events. For example, the events "switch computer on" and "display DTM on computer screen" are causally related events because the second event cannot happen before the first one. But "switch computer on" and "answer the telephone call" are concurrent events because they have nothing to do with each other.

Every event is unique in space and time, but it is useful to collect them into **event classes** and give a characteristic name to each of them in order to emphasize common structure and behavior. The events "Mr. Sawyer looks at DTM" and "Ms. Smith looks at DTM" are both instances of the same event class *Person looks at DTM*. If an event class has attributes, then the event is written as an instance of this class with its attributes added in parenthesis, e.g. "Mr. Sawyer looks at DTM (scale, viewAngle)".

A **scenario** is defined as a sequence of events that may happen within a system domain. It may be a real record of historical events, or include a thought experiment of possible events. The length or scope of a scenario can be different, depending on whether all events of a system are selected, or only specific ones. The normally used form of a scenario is a pure textual description of the events taking place in their actual sequence; it does neither mention the state of objects nor the causes of events. The following example is a scenario that a computer user might experience without a power protection system (Fig. 2.16).

User switches power button on
Power supply goes on
Computer starts to work
Power supply goes off
User switches power button off
Power supply goes on
User switches power button on
Power supply goes on
Computer does not work
User presses reset button
Computer does not work
User switches power button off
Power supply goes off
User switches power button on
Power supply goes on
Computer does not work
User hits computer

Fig. 2.16. A possible scenario for a computer user without a power protection system

A scenario can be the first step to design a state-transition diagram. Before doing this, it is often helpful to write down an event trace. An **event trace** is defined as an augmented scenario where the sending and receiving objects are shown as vertical lines, and each event as a horizontal arrow pointing from the sender object to the receiver object. The event trace starts at the top with the first event, and as time increases, all other events follow one after the other. Because the timing is not important, all events are written with an equal distance to each other in the event trace. Fig. 2.17 shows an event trace for the scenario described in Fig. 2.16.

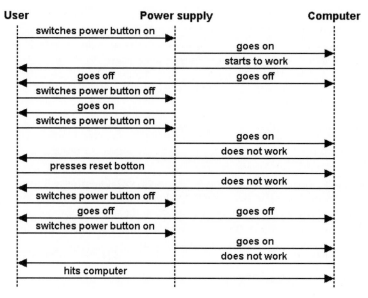

Fig. 2.17. A possible event trace for a computer user

A **state-transition diagram** is a graphical representation of the different states of one object responding to incoming events. A change of state due to an event is defined as a **transition**, also called the "firing" of an event. After receiving an event, the next state of an object depends on both its present state and the event itself. The OMT notation for a state-transition diagram contains rounded boxes for the different states of an object connected by arrows starting at the receiving state box and pointing to the target state box. An arrow is labeled with the name of the corresponding event causing the transition. Solid black circles to mark its start and end can supplement the beginning and the end of a state-transition diagram. Because the state-transition diagram deals with all different states of an object, it is a graph describing a sequence of states caused by a sequence of events. If a graph is more complex, an event sequence is a path through the graph. If an event has no effect on the state of an object, it is ignored and the object keeps its present state. Fig. 2.18 and Fig. 2.19 contain three different state-transition diagrams for user, power supply, and computer according to the scenario introduced above.

Note that the event "User presses reset button" was not included in Fig. 2.19 because it does not change the state "On / Not O.K." of the computer object in the considered scenario. In certain situations it is possible that one single event may not result in one unique transition, but in different ones resulting in different object states. Depending on some circumstances, an event may cause a multitude of transitions and object states. A **condition** is defined as a Boolean statement about object attributes that alter the transition and states of objects. Conditions are often used as guards on transitions, and then also called guard conditions. If the Boolean statement is true, one transition is executed resulting in a certain object state. If the statement is false, another transition takes place giving the object another state. In state-transition diagrams conditions are shown in brackets following the event name. Fig. 2.20 displays a state-transition diagram for computer, which is a changed version of Fig. 2.19 including a condition at the end.

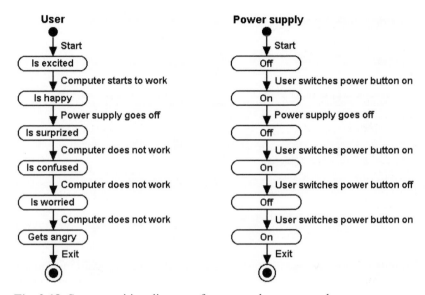

Fig. 2.18. State-transition diagrams for user and power supply

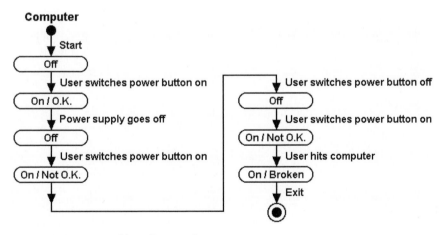

Fig. 2.19. State-transition diagram for computer

In Fig. 2.20 the state of the computer object depends on whether the user hits the right or the wrong spot. Depending on this condition the state of the computer object is "On / O.K." or "On / Broken". Conditions can be valid for some period of time. In this aspect, they are different from events, which have no time duration. State-transition diagrams can describe one-shot life cycles like in Fig. 2.18 to Fig. 2.20, or continuous loops as in Fig.1.6. They represent the behavior of a class of objects. Because all objects of one class share per definition the same behavior, they also share the same state-transition diagram.

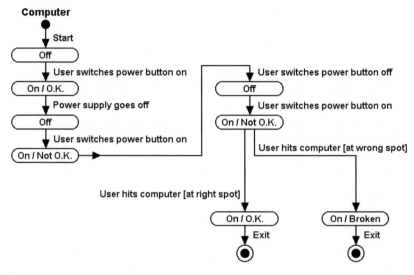

Fig. 2.20. State-transition diagram for computer with a condition

The state-transition diagrams introduced so far are able to represent the behavior of single classes of objects in terms of different events and states. When a change of state happens, it is natural to connect this transition with some type of operation that causes the event to happen. In the OMT, one distinguishes between two different types of operations. (i) An **activity** is defined as an operation that needs some time to complete. Activities are rooted in one state of an object. They may include continuous operations like looking at a computer screen, or sequential operations that stop by themselves after some time, such as booting a computer system. (ii) An **action** is defined as an instantaneous operation associated with an event. "User hits computer", for example, is an action in Fig. 2.19 or Fig. 2.20. An action has no time duration in comparison to the time period needed to run through the state-transition diagram from start to exit. The notation for activities is "do: activity" written into a state box. It starts on entry to the state and stops on exit for a continuous operation. In case of a sequential operation, it starts on entry to the state and stops when it is finished or interrupted by another event. The notation for an action is a slash ("/") after the event name followed by the action name or its description. If necessary, a state box can be further supplemented by entry, exit, or internal event actions executed on entry to the state, on exit, or when the internal event takes place. Fig. 2.21 summarizes the different notations, using a less rounded symbol for a state box, which can help to save space in a packed state-transition diagram.

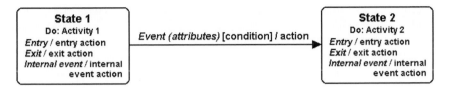

Fig. 2.21. Summary of notation for state-transition diagrams

Summarizing, a state name is given in bold characters inside of a rounded box in Fig. 2.21. Events are written in italics to be able to distinguish them better from activities and actions. They are separated from each other by a slash ("/"). A transition arrow is labeled by at least the event name followed by attributes in parentheses, a condition in brackets, and an action in the end. Attributes, condition, and action are all optional. Fig. 2.22 illustrates the notation in Fig. 2.21 for an input dialog box in a windows program where "*Internal Event*" is shortened to "*Event*".

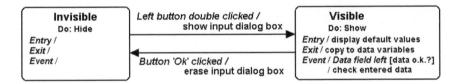

Fig. 2.22. State-transition diagram for an input dialog box

After the left mouse button is double clicked, the input dialog box leaves its state "Invisible" and becomes "Visible". When this happens, the entry action is performed which is to display the default or already entered values of each data input field in the dialog box. During the input process, when the internal event "*Data field left*" occurs, an internal action "check the entered data" with the condition "data o.k.?" executes. If any entered data is not acceptable, a warning box pops up. After finishing the input process, the "Ok" button is pressed. Then the exit action is executed by which the entered data values are copied to their corresponding data variables. After completing this, the transition is reversed and the input dialog box becomes again "Invisible".

The dynamic model of the OMT consists of other more advanced topics, such as nested state-transition diagrams, state or event generalization. They will be considered in later parts of the book, if the need arises. For example, nested state-transition diagrams are used to describe the states of more complex systems where unstructured state-transition diagrams are not sufficient anymore.

2.4.3 Functional model

The functional model completes the object and dynamic model of the OMT like the third leg of a tripod. Its focus is on the transformations of values within the system, occurring in functions, mappings, constraints, and functional dependencies, for example. It captures what a system is doing computational wise, but it does not look into timing aspects or implementation details. The functional model is shown in a data flow diagram to represent the transformations of data within the system. They display the relation between data sources, processes, and data stores (where data comes from, what it is transformed into, where it goes to, and where it is stored).

The functional model deals with the flow of data between stores and processes but not its control, although one is sometimes tempted to include control information into a data flow diagram. However, this is not correct. In the OMT the dynamic model only describes control issues, whereas the functional model considers just the flow of data and its various aspects. It only reveals what functional paths are available in the system. It does not show what way the system will actual take. The decision on this belongs to the system control and is taken in the dynamic model.

Data flow diagrams are already well known from structured modeling techniques. They are based on (Rumbaugh et al. 1991):
- Processes that transform data,
- Data flows that move data,
- Actors that produce and consume data, and
- Data stores that store data.

A **process** is defined as an operation that transforms data inputs into data outputs. For example, to compute a DTM using three-dimensional Cartesian coordinates x,y,z is a process. The coordinates are the data input, the DTM is the data output. The notation for a process in a data flow diagram is an ellipse containing the process name or its description. As part of the data flow, data inputs are drawn as arrows pointing into the ellipse, data outputs as arrows coming out of the ellipse. Labels are attached to the arrows carrying the names of data inputs and outputs, or their descriptions. It is possible that one process has more than one data input or output. In such a case, more than one arrow goes into or leaves an ellipse. Fig. 2.23 illustrates the introduced notation for the DTM example.

Fig. 2.23. Process notation for the DTM example

A **data flow** is defined as a channel or pipe that data passes when moving from a producer to a consumer. A producer or consumer may be another process or may belong to the operations of a class of objects. The data values remain unchanged by a data flow, but splitting or merging parts of data can affect the data components, for example. Consider, for instance, the three-dimensional Cartesian coordinates x,y,z mentioned before. At some stage of the DTM it is necessary to separate the z coordinate from x and y to produce a three-dimensional view of the terrain. The corresponding data flow is shown in the right-hand part of Fig. 2.24.

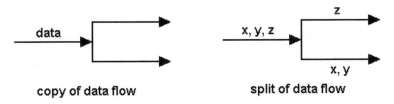

copy of data flow   split of data flow

Fig. 2.24. Separated data flows

In the left-hand flow of Fig. 2.24, the same data is identically copied into two flows without any separation. An **actor** is defined as an active object, which produces or consumes data. They normally terminate a flow of data, for example, from a user as the source of x,y,z coordinates to the computer screen where the data flow ends by displaying the DTM. Thus, actors are also called "terminators". The notation for an actor in a data flow diagram is a rectangle labeled with the actor's name. A rectangle was chosen as the actor's symbol to indicate that an actor is an object. In Fig. 2.25 a context diagram illustrates the introduced notation, which is the highest-level data flow diagram in the OMT with only one process.

Fig. 2.25. Context diagram with actors as parts of a DTM

In Fig. 2.25 the process is to "compute a DTM" with the data input "x,y,z" and the data output "DTM". "User" and "Computer screen" are the actors at both ends of the data flow. A **data store** is defined as a passive object representing a place to hold data for later access. Data stores can keep data either temporarily like in a buffer, or more permanently on a storage device, for example, a floppy disk. The notations for a data store in a data flow diagram are two parallel lines with the name of the data store written in the middle. Using this notation, the context diagram in Fig. 2.25 can be decomposed as shown in Fig. 2.26.

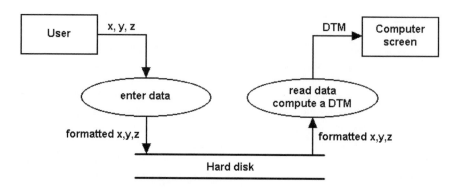

Fig. 2.26. DTM data flow diagram with a data store

In this diagram the process "compute a DTM" in Fig. 2.25 is replaced and re-fined by adding a process and a data store. In the beginning, the user enters x,y,z coordinates which are stored in some format on the hard disk as the data store. Then the formatted coordinates are read from the hard disk and used to compute the DTM before passed on to the computer screen.

Every process in the functional model is finally modeled by an operation as an integral part of a class of objects. For these operations the functional model is the tool to design their data inputs and outputs together with their functionality. For example, a class for data management includes operations to create, delete, read, write, and modify data.

2.5 Coad's object-oriented method

In this paragraph Coad's object-oriented method will be considered in more detail. It is an extension and refinement of the method of Coad & Yourdon (1991) men-tioned before. The main characteristic of Coad's method is that it models a system using a more complete picture and not different projections of the same system as the other techniques do. Like a CAD program it keeps the viewpoint and applies different layers, each layer adding more detail to the whole picture like overlap-ping transparencies. Coad's method uses at least the following layers in a sequence from top (more general) to bottom (more detailed):
- Class/Object layer,
- Attribute layer, and
- Service layer.

For more complex systems other layers (subject and structure layer) can be added. In order to increase the stability and reusability of their results, Coad et al. (1995) recommend decomposing the whole system domain into four components:

- Problem domain,
- Human interaction (with computers),
- Data management, and
- System interaction (between different hardware).

Problem domain (PD) components contain classes/objects that directly address the problem under consideration. They should be as independent as possible from implementation issues and the other components, which allows an easier exchange with another PD component, thus enhancing stability and reusability.

Human interaction (HI) components contain classes/objects that create and manage interfaces between human users and PD components. They are, for example, responsible for graphical user interfaces, which will also incorporate more and more interaction with human voice and touch in the near future.

Data management (DM) components contain classes/objects that create and manage interfaces between PD components and database or file management systems. They are responsible for data storage, update, and retrieval, and support features like searching, inserting, deleting, or sorting of data, which are much related to PD and HI components.

System interaction (SI) components contain classes/objects that create and manage interfaces between PD components and other hardware systems or devices, e.g. in a network. This requires, for example, a common communication protocol or the management of different driver software.

Coad's method is based on four major activities, each of which may be performed with each of the system domain components PD, HI, DM, and SI described before (Coad et al. 1995):

- **Activity 1:** Identify the system's purposes and behavior.
- **Activity 2:** Identify the classes/objects of each system domain component and organize them by applying object patterns, which are defined as templates of classes/objects with stereotyped responsibilities and interactions. For example, whole-to-part connections are object patterns.
- **Activity 3:** Identify responsibilities for each class/object:
  - What the class/object knows (its attributes),
  - Who the class/object knows (its connections), and
  - What the class/object does (its services).
- **Activity 4:** Work out the dynamics of each of the system domain components (PD, HI, DM, SI), using scenarios, which are sequences of events of class/object interactions needed to fulfill a responsibility.

Coad's method uses a notation consisting of seven symbols shown in Fig.2.27.

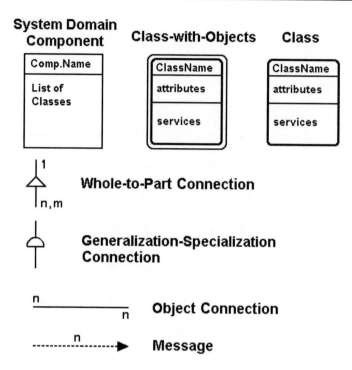

Fig. 2.27. Notation symbols of Coad's method

The symbols for **class-with-objects, class,** and **whole-to-part connection** were already explained in paragraph 2.2. The symbol for the system domain component is a rectangular box with the component's name and a list of classes inside of the box. The **generalization-specialization connection** links a general class-with-objects or class to a related specialized class-with-objects or class. For example, the general class *Computer* can have a specialization in form of a class *Laptop*. The symbol of a generalization-specialization connection is directional; the semi-circle points in direction of the general class in the same way the triangle of the whole-to-part connection points to the whole class. Sometimes objects are linked together without a whole-to-part or a generalization-specialization connection if, for example, their attributes and/or services are needed by the other object to fulfill its responsibilities. In these cases an **object connection** exists between the related objects. The labels "n" at both ends of the object connection represent the number of involved objects on the left-hand or right-hand side of the connection. The last symbol of Coad's notation in Fig. 2.27 is an arrow representing a message. A **message** is defined as a request passed from a sender (a class/object) to a receiver (another class/object) that should provide a service. The arrow is drawn from the sending class/object in the direction of the class/object that receives the message. The label "n" next to the arrow means that a message with or without parameters is sent to another service in the same class or a different class.

A small example with an introduction to Coad's notation was already given in paragraph 2.2. This example is further extended as follows to demonstrate how the different symbols of Coad's notation can be used.

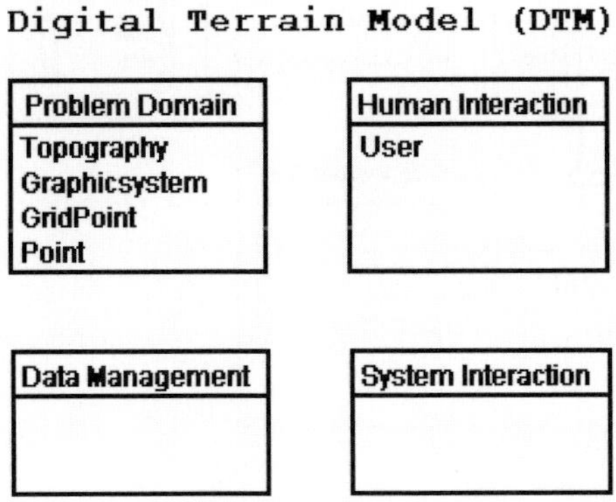

Digital Terrain Model (DTM)

Problem Domain
Topography
Graphicsystem
GridPoint
Point

Human Interaction
User

Data Management

System Interaction

Fig. 2.28. DTM object model

Fig. 2.28 shows the four system domain components PD, HI, DM, and SI with their associated classes, whereby the DM and SI components were left empty to keep the model as simple as possible. This object model can be refined as shown in Fig. 2.29 where the DM and SI components were not included.

In this object model class *GridPoint* is a specialization of the general class *Point*. Apart from some attributes not further specified here, class *User* of the human interaction component contains two services "start" and "exit" to begin and exit the model. To complete the model a connection is needed between class *User* and *Topography* across different components. Such a connection could be included in Fig. 2.29 but generally, an object model gets too overloaded with information if all these connections are inserted. Therefore, it is better not to do this and show class connections across components inside of scenarios. A **scenario** in Coad's method is defined as a time-ordered sequence of object interactions needed to fulfill a specific service responsibility. As it takes several sequential steps to repair a computer, for example, these steps would make up the steps for the scenario "repairComputer". For the DTM object model in Fig. 2.29 a scenario diagram "drawTopography" can be described as shown in Fig. 2.30, whereby the sequence of steps starts at the top and finishes at the bottom.

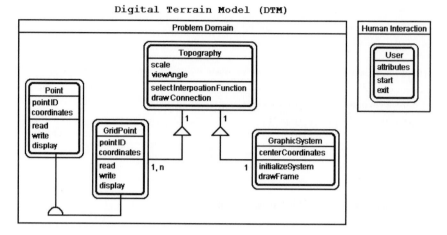

Fig. 2.29. More detailed DTM object model

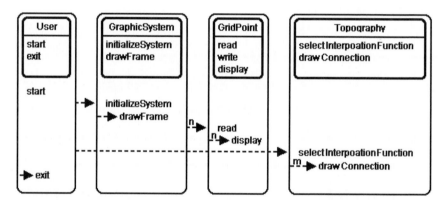

Fig. 2.30. Scenario diagram of "drawTopography" for the DTM object model

In Fig. 2.30, class *User* who sends a message to class *GraphicSystem* to initialize the system and draw the coordinate frame starts this scenario. Then *GraphicSystem* gives a message n times to class *GridPoint* to read pointIDs, coordinates, and display them. Then class *User* chooses an interpolation function by invoking the corresponding service of class *Topography* called "selectInterpolationFunction". Finally, class *Topography* draws a connection between different grid points m times, before class *User* exits the scenario.

Because scenarios show how services of one class are related to services of the same or other classes, it is obvious that scenario diagrams can become very large because they merge object interaction, object messaging, and state-transition concepts. A scenario consists of at least three parts:
• The sender class and service,

- The receiver class and service, and
- The parameters and number of service invocations.

Parameters or IDs of service invocations may be optional, but a scenario has to know at least the sending and receiving class as well as the wanted service. **Parameters** are defined in Coad's method as data belonging to a message. If the sender class needs to know the outcome of the invoked service, a message with this information is passed back from the receiver class. For example, a very common return message is to send back a Boolean value to the sending class, indicating the service was executed successfully (true) or without success (false).

2.6 The unified modeling language (UML)

2.6.1 Introduction

Booch, Rumbaugh & Jacobson (1998) jointly developed the unified modeling language (UML) as an evolution of the Booch, OMT, OOSE and some other methods. It was started by Grady Booch and Jim Rumbaugh in 1994 to combine the Booch method and the OMT. Later Jacobson (1992) joined them who created the "Object-Oriented Software Engineering" (OOSE) method, and who was the first to propose use-cases. Thus, the UML is basically a synthesis of Booch's method and the OMT with other concepts and diagrams being added in time, as reflected in the table below where state-transition diagrams are shortened to state diagrams.

Table 2.2. A comparison of diagrams with an identical or similar meaning

Booch's method	OMT	UML (version 1.1)
Class / Object diagram	Class diagram	Class / Object diagram
State diagram	State diagram	State diagram
Scenario diagram (= Sequence diagram)	Scenario diagram (= Sequence diagram)	Scenario diagrams: Sequence diagram Collaboration diagram
Module diagram	---	Component diagram
Process diagram	---	Deployment diagram
---	---	Use-case diagram
---	---	Activity diagram

As can be seen in this table, the UML retains the general structure of Booch's method, recommending using static and dynamic views of a logical model and a physical model of the problem/system under consideration. The UML notation, which is close to the OMT notation in their common parts, enables the user to create and refine these different views within an overall model, representing the problem domain and the corresponding software system. The **goal of the UML** is to provide a common standard for specifying, visualizing, constructing and documenting OO models. It became the industry standard in 1997, acknowledged by the "Object Management Group" (OMG), an industry standards body. The UML prescribes a standard set of diagrams and notations for modeling OO systems, and describes the underlying semantics of what these diagrams and symbols mean. Whereas there has been to this point many notations and methods used for object-oriented design, now there is a single notation for builders to learn. With an increasing number of complex systems existing, the UML is being further extended and updated. The latest version and specifications can be found at http://www.omg.org/uml/.

The UML version 1.1, for example, has an overwhelming number of 233 concepts (84 basic and 149 diagram concepts). This is because the UML is process-independent and by its nature, complexity is not simple to model. But who can read, understand and use such a huge amount of concepts? At this point many people give up in frustration because basically, the UML is a language for modeling but it does not guide a developer in how to build a model, or what building plan to follow. This is the reason why material like this text are compiled and predecessors of the UML included: to organize complexity and help the reader to make necessary simplifications.

The UML provides graphical and textual representations for
- Class models,
- Component models,
- Deployment models, and
- Use-case models.

Corresponding diagrams represent each of these models. Apart from class and component models already covered by Booch's method and the OMT, the UML allows to set up deployment models and diagrams (which are similar to process diagrams in Booch's method) and use-case models as new extensions. Another new part of the UML are collaboration diagrams as another type of interaction diagrams which show how two or more objects that participate in a client/server relation work together in order to provide a service. An example for a collaboration diagram is given in Fig. 2.31.

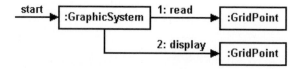

Fig. 2.31. Example for a collaboration diagram

In Fig. 2.31, an object of class *GraphicSystem* passes two messages for reading and displaying data to an object of class *GridPoint*. Compared to sequence diagrams collaboration diagrams can be more advantageous because they can express more contextual information and are smaller in space. However, either diagram type can show similar constructs.

Component models were not explicitly mentioned in Booch's method or the OMT, but covered in principle by the module diagrams in Booch's method. In the UML, components are defined as physical units of source code and executable units that are assembled into a model implementation. Classes are assigned to components as reusable building blocks. These components form the foundation for interchangeable architectures.

Deployment models allow determining how an application can be mapped to a distributed computer network. The UML makes it possible to model different network topologies, such as client/server, three-tired architectures or Internet/Intranet networks. In the UML, the topology of nodes within the network is described, how these nodes are connected, and how the solution is partitioned and distributed across this network.

Use-case models have become a very essential part of requirements analysis in many industries. A use-case can be defined as a collection of related scenarios, each of which represents a prototypical behavior of the problem/system to be modeled. Use-cases allow defining rules and tasks in a business or manufacturing process, for example. They determine how these tasks can be supported by an application system. Use-cases capture the requirements of the problem/system under consideration and allow mapping them to a class model. Thus, use-case models help to bridge the gap between domain analysts, users, and application developers. Use-cases are generally the starting point of object-oriented analysis with UML. Use-case models consist of actors and use-cases. Actors represent users and other systems that interact with the system to be constructed. They are drawn as stick figures. They actually represent a type of user, not an instance of a user class. Use-cases represent the behavior of the system and what it goes through in response to stimuli from an actor. They are drawn as ellipses and connected to actors through arrows. An example for a use-case diagram is given in the figure below.

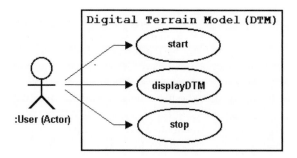

Fig. 2.32. Use-case diagram for a Digital Terrain Model (DTM)

2.6.2 UML diagrams and how to use them

As listed in Table 2.2, the UML 1.1 version contains **nine diagrams**:
- Class diagrams for modeling the static structure of classes,
- Object diagrams for modeling the static structure of objects,
- State diagrams for modeling the state behavior of objects,
- Sequence diagrams for modeling message passing between objects,
- Collaboration diagrams for modeling interactions between objects,
- Use-case diagrams for modeling all processes inside of a problem/system,
- Activity diagrams for modeling the behavior of uses cases and objects,
- Component diagrams for modeling components, and
- Deployment diagrams for modeling the distribution of the model.

When building an OO model, these diagrams are generally used in a certain order as given below that originates from the way a builder performs a job. Generally, a builder starts with the requirements and continues with the static framework of a construction before moving to its dynamic behavior, which may require to redo the already existing static framework, and so on. Such an iterative working scheme is also called round-trip engineering described in paragraph 3.1. Of course, such an order is not fixed and represents just a guideline to help the reader to better grasp the role and meaning of each diagram.

1. Use-case diagrams,
2. Activity diagrams,
3. Component diagrams,
4. Class diagrams,
5. Sequence diagrams,
6. Collaboration diagrams,
7. Object diagrams,
8. State diagrams, and
9. Deployment diagrams.

How to use each diagram is briefly described below (Popkin 2001).

1. Use-case diagrams

The UML is a notation, not a method. It does not prescribe a work plan for modeling a system. However, because the UML includes use-case diagrams, it is considered to lend itself to a problem-centric, use-case driven modeling approach. Use-cases are generally viewed as the simplest and most effective technique for capturing system requirements from a user's perspective. Use-cases are used to model how a system currently works, or how the users wish it to work. It is not really an object-oriented approach; it is really a form of process modeling. It is, however, an excellent way to lead into OO system analysis. Use-cases are generally the starting point of OO analysis with UML. They provide entry points into analyzing the requirements of the system under consideration, and the problem that needs to be solved. Processes are described within these use-cases by a textual description or a sequence of steps performed. Each use-case is documented by a description of the scenario. The description can be written in textual form or in a step-by-step format. Each use-case can also be defined by other properties, such as the pre- and post-conditions of the scenario - conditions that exist before the scenario begins, and conditions that exist after the scenario completes. Once the system's behavior is captured in this way, use-case diagrams are examined and amplified to show what objects interrelate to make this behavior happen.

2. Activity diagrams
Activity diagrams are multi-purpose process flow diagrams that are used to model the behavior of a system. They can be used to model a use-case, or a class, or a complicated operation. An activity diagram is similar to a flow chart; the one key difference is that activity diagrams can show parallel processing. This is important when using activity diagrams to model system processes, some of which can be performed in parallel, and for modeling multiple threads in concurrent programs. Activity diagrams provide a graphical tool to model the process of use-cases. They can be used in addition to, or in place of, a textual description of the use-case, or a listing of the steps of the use-case. A textual description, code, or another activity diagram can detail the activity further. When modeling the behavior of a class, a UML state diagram is normally used to model situations where asynchronous events occur. The activity diagram is used when all or most of the events represent the completion of internally generated actions. You should assign activities to classes before completing an activity diagram.

3. Class diagrams
As the objects are found, they can be grouped by type and classified in a class diagram. It is the class diagram that becomes the central analysis diagram of the object-oriented design, and one that shows the static structure of the system. The class diagram can be divided into different groups, e.g. for processing or data managing. It shows the classes involved with the user-interface, system logic, and data storage, for example. Component diagrams are used to group classes into components or modules. Overall hardware distribution of the system is modeled using deployment diagrams.

4. Component diagrams
Component diagrams allow to model and structure system domains, use-cases, classes, or components by using packages. In the UML, a **package** is the universal item to group elements, enabling builders to subdivide and categorize systems. Packages can be used on every level, from the highest level, where they are used to subdivide the system into domains, to the lowest level, where they are applied to group individual use-cases, classes, or components.

5. Sequence diagrams
Sequence diagrams are one of the most effective diagrams to model object interactions in a system. A sequence diagram is modeled for every use-case. Whereas the use-case diagram enables modeling of a processing view of a sce-

nario, the sequence diagram contains implementation details of the scenario, including the objects and classes that are used to implement the scenario, and messages passed between the objects. Typically one examines the description of the use-case to determine what objects are necessary to implement the scenario. If you have modeled the description of the use-case as a sequence of steps, then you can "walk through" the steps to discover what objects are necessary for the steps to occur. A sequence diagram shows objects involved in the scenario by vertical dashed lines, and messages passed between the objects as horizontal vectors. The messages are drawn chronologically from the top of the diagram to the bottom; the horizontal spacing of objects is arbitrary.

6. Collaboration diagrams

Collaboration diagrams represent an alternative to sequence diagrams for modeling interactions between objects in the system. Whereas in the sequence diagram the focus is on the chronological sequence of the scenario being modeled, in the collaboration diagram the focus is on understanding all of the effects on a given object during a scenario. Objects are connected by links, each link representing an instance of an association between the classes involved. An association is also a class for defining a semantic relationship between classifiers, such as classes. The link shows messages sent between the objects, the type of message passed (synchronous, asynchronous, simple, balking, and time-out), and the visibility of objects.

7. Object diagrams

Object diagrams encompass objects and their relationships at a point in time. An object diagram may be considered as a special case of a class diagram or a collaboration diagram.

8. State-transition or shorter State diagrams

While sequence and collaboration diagrams model dynamic actions between groups of objects in a system, the state diagram is used to model the dynamic behavior of a particular object, or class of objects. A state diagram is modeled for all classes deemed to have significant dynamic behavior. In it, you model the sequence of states that an object of the class goes through during its life in response to received stimuli, together with its own responses and actions. For example, an object's behavior is modeled in terms of what state it is in initially, and what state it transitions to when a particular event is received. You also model what actions an object performs while in a certain state. States represent the conditions of objects at certain points in time. Events represent incidents that cause objects to move from one state to another. Transition lines depict the movement from one state to another. Each transition line is labeled with the event that causes the transition. Actions occur when an object arrives in a state.

9. Deployment diagrams

Deployment diagrams are used to model the configuration of run-time processing elements and the software components, processes, and objects that live on them. In the deployment diagram, you start by modeling the physical nodes and the communication associations that exist between them. For each node, you can indicate what component instances live or run on the node. You can also model the objects that are contained within the component. Deployment diagrams are used to model only components that exist as run-time entities; they are not used to model compile-time only or link-time only components. You can also model components that migrate from node to node or objects that migrate from component to component using a dependency relationship.

2.6.3 UML 1.1 extensions

Built-in extensibility mechanisms enable UML to be a somewhat open specification that can cover aspects of modeling not specified in the version 1.1 document (Popkin 2001). These mechanisms enable UML notation and semantics to be expanded by stereotypes, business modeling extensions, and formal languages, for example.

Stereotypes
Stereotype is the most widely used built-in extensibility mechanism within UML. A stereotype represents a usage distinction. It can be applied to any modeling element, including classes, packages, inheritance relationships, etc. For example, a class with stereotype <<actor>> is a class used as an external agent in the system modeling. A template class is modeled as a class with stereotype <<parameterized>>, meaning it contains parameters.

Business Modeling Extensions
A separate document within UML specification calls out specific class and association stereotypes that extend UML to cover business-modeling concepts. This includes stereotyping a class as an actor, a worker (both internal and case), or an entity, and stereotyping an association as a simple communication, or a subscription between a source and a target.

Object Constraint Language (OCL)
A picture can only describe so many words. Similarly, a graphical model can only describe a certain amount of behavior, after which it is necessary to fill in additional details with words. Describing something with words, however, almost always results in ambiguities; i.e., "what did he mean when he wrote that?" The Object Constraint Language (OCL) is incorporated into UML as a standard for specifying additional details, or precise constraints on the structure of the models. Developed within the IBM Insurance Division as a business modeling language, the OCL is a formal language designed to be easy to read and write. OCL is more formal than natural language, but not as precise as a programming language - it cannot be used to write program logic or flow control. Since OCL is a language for pure expression, its statements are guaranteed to be without side effects - they simply deliver a value and can never change the state of the system.

2.6.4 Comparing UML to other modeling languages

It should be made clear that the Unified Modeling Language is not a radical departure from Booch, OMT, or OOSE, but rather the legitimate successor to all three (UML Summary v.1.1, http://www.omg.org/uml/). This means that if you are a Booch, OMT, or OOSE user today, your training, experience, and tools will be preserved, because the UML Language is a natural evolutionary step. The UML will be equally easy to adopt for users of many other methods, but their authors must decide for themselves whether to embrace the UML concepts and notation underneath their methods.

The UML is more expressive yet cleaner and more uniform than Booch, OMT, OOSE, and other methods. This means that there is value in moving to the UML, because it will allow projects to model things they could not have done before. Users of most other methods and modeling languages will gain value by moving to the UML, since it removes the unnecessary differences in notation and terminology that obscure the underlying similarities of most of these approaches.

With respect to other visual modeling languages, including entity-relationship modeling, flow charts, and state-driven languages, the UML should provide improved expressiveness and holistic integrity. Users of existing methods will experience slight changes in notation, but this should not take much relearning and will bring a clarification of the underlying semantics. If the unification goals have been achieved, UML will be an obvious choice when beginning new projects, especially as the availability of tools, books, and training becomes widespread. Many visual modeling tools support existing notations, such as Booch, OMT, OOSE, or others, as views of an underlying model; when these tools add support for UML (as some already have) users will enjoy the benefit of switching their current models to the UML notation without loss of information.

Existing users of any OO method can expect a fairly quick learning curve to achieve the same expressiveness, as they previous knew. One can quickly learn and use the basics productively. More advanced techniques, such as the use of stereotypes and properties, will require some study, since they enable very expressive and precise models, needed only when the problem at hand requires them.

2.6.5 New features of the UML

The goals of the unification efforts to create the UML were to keep it simple, to cast away elements of existing Booch, OMT, and OOSE that didn't work in practice, to add elements from other methods that were more effective, and to invent new only when an existing solution was not available (UML Summary v.1.1, http://www.omg.org/uml/). Because the UML authors were in effect designing a language (albeit a graphical one), they had to strike a proper balance between minimalism (everything is text and boxes) and over-engineering (having an icon for

for every conceivable modeling element). To that end, they were very careful about adding new things, because they didn't want to make the UML unnecessarily complex. Along the way, however, some things were found that were advantageous to add because they have proven useful in practice in other modeling. There are several new concepts that are included in UML, including extensibility mechanisms: stereotypes, tagged values, and constraints; threads and processes; distribution and concurrency (e.g. for modeling ActiveX/DCOM and CORBA); patterns/collaborations; activity diagrams (for business process modeling); refinement (to handle relationships between levels of abstraction); interfaces and components; and a constraint language.

Many of these ideas were present in various individual methods and theories but UML brings them together into a coherent whole. In addition to these major changes, there are many other localized improvements over the Booch, OMT, and OOSE semantics and notation. The UML is an evolution from Booch, OMT, OOSE, several other object-oriented methods, and many other sources. These various sources incorporated many different elements from many authors, including non-OO influences. The UML notation is a melding of graphical syntax from various sources, with a number of symbols removed (because they were confusing, superfluous, or little used) and with a few new symbols added. The ideas in the UML come from the community of ideas developed by many different people in the object-oriented field. The UML developers did not invent most of these ideas; rather, their role was to select and integrate the best ideas from OO and computer-science practices. The actual genealogy of the notation and underlying detailed semantics is complicated, so it is discussed here only to provide context, not to represent precise history.

Use-case diagrams are similar in appearance to those in OOSE. Class diagrams are a melding of OMT, Booch, and class diagrams of most other OO methods. Extensions (e.g., stereotypes and their corresponding icons) can be defined for various diagrams to support other modeling styles. Stereotypes, constraints, and tagged values are concepts added in UML that did not previously exist in the major modeling languages. State diagrams are substantially based on the state charts of David Harel with minor modifications. The Activity diagram, which shares much of the same underlying semantics, is similar to the work flow diagrams developed by many sources including many pre-OO sources. Oracle and Jim Odell were instrumental in incorporating Activity Diagrams into UML. Sequence diagrams were found in a variety of OO methods under a variety of names (interaction, message trace, and event trace) and date to pre-OO days. Collaboration diagrams were adapted from Booch (object diagram), Fusion (object interaction graph), and a number of other sources. Collaborations are now first-class modeling entities, and often form the basis of patterns. The implementation diagrams (component and deployment diagrams) are derived from Booch's module and process diagrams, but they are now component-centered, rather than module-centered and are far better interconnected. Stereotypes are one of the extension mechanisms and extend the semantics of the meta model. User-defined icons can be associated with given stereotypes for tailoring the UML to specific processes.

Object Constraint Language is used by UML to specify the semantics and is provided as a language for expressions during modeling. OCL is an expression language having its root in the Syntropy method and has been influenced by expression languages in other methods like Catalysis. The informal navigation from OMT has the same intent, where OCL is formalized and more extensive. Each of these concepts has further predecessors and many other influences. It is realized that any brief list of influences is incomplete and that the UML is the product of a long history of ideas in the computer science area.

2.6.6 New developments affecting the UML

Some new developments have started with the goal to improve the overall versatility, quality, and productivity of modeling and implementation methods. One development is the growing incorporation of more intelligent and self-learning components in the modeling and implementation process. Another development are broader and more versatile modeling approaches. Their goal is always to catch more functionality and reusability in a model that cannot be satisfactorily solved by just using object-oriented methods. Two approaches have gained more attention for combining design elements in principled and proven ways: architectures and patterns.

Architectures can be defined in the context considered here as high-level models of a problem/system (Monroe et al. 1997). A modeling approach using architectures has its focus more on the solution and less on the problem/system. It provides a set of high-level abstractions for finding a solution to the problem/system under target. These abstractions are expressed by groups of architectural idioms, also called "architectural styles". Each architectural style has its own notation used to describe the structure and behavior of the problem/system. Object-orientated modeling is just one type of different architectural styles. Other styles are *procedural* based on a call-return model, *dataflow, event-driven, layered* where the system functionality is distributed among a hierarchy of layers, client/server or state machine, which describes the system behavior by a sets of states together with a set of services linked to each state. Key topics of every architecture are, for example, the overall system behavior, scaling and portability of the designed model, and global system properties like processing rates, end-to-end capacities, or overall performance of the complete model.

The idea to use **patterns**, also called design patterns, for modeling a problem/system is based on the fact that the same patterns can be found across various applications, but looking differently from an outside perspective (Kerth & Cunningham 1997). These patterns can be extracted, described, and reused in different models, adopting each of them to the specific problem/system under consideration. Thus, design patterns provide common solutions to related problems, which helps to simplify each model, and they allow to rise a model above the component level by including information on the system level. This information includes human elements and characteristics in order to capture the wholeness of the system. For example, to increase a model's user-friendliness basic human comforts need to be considered and incorporated. Patterns will be described in more detail in the next paragraph.

These ideas and approaches lead immediately to the fields of artificial intelligence (AI) and expert systems. One **goal of AI** research is to analyze human behavior regarding perception, comprehension and decision making in order to reproduce and simulate them on computers. **Expert systems** are the first true applications that come closer to reach this goal. The relevance of AI systems is growing because the human demands for more intelligent applications seem to have no limit. The underlying driving force is the human ability to imagine improved solutions that will always exceed the human ability to built them. Thus, the desires of human imagination drive the technical development of the future (Booch 1996). In the present age of technological advances, for example, a whole variety of new systems are being developed and constructed following this trend. It includes hand-held communication devices that work at any place of the world, miniature systems for applications in environmental sciences and medicine, or robots that are directed by image and speech recognition techniques. Because these systems are becoming also more complex, object-oriented methods play a key part in their development, offering the following advantages (Booch 1996): better time to market, improved quality and increased reuse.

2.7 Objects and patterns

2.7.1 Introduction

Experience in OO modeling has helped to build up a repertoire of both general principles and idiomatic solutions that represent guidelines in the building of models (Larman 1998). These principles and idioms, if codified in a structured format describing the problem and solution together with a given name, may be called **patterns**. For example, assigning a responsibility to the most suitable class could be the pattern for an expert.

In object technology, a pattern is **defined** as a named description of a problem and solution that can be applied to new contexts, with advice on how to apply it in varying circumstances. The formal notation of patterns originated with the architectural patterns of Alexander (1977) who wrote: "Each pattern describes a problem which occurs over and over again in our environment, and then describes the core of the solution to that problem, in such a way that you can use this solution a million times over, without ever doing it the same way twice".

In time, a lot of different patterns have evolved, e.g. design patterns, concurrency patterns, **General Responsibility Assignment Software Patterns** or GRASP patterns, GUI design patterns, coding patterns, optimization patterns or testing patterns (Grand 1998). The more common ones will be further described here after, and then illustrated in part II of this book with several program examples.

2.7.2 Design patterns

The application of patterns to software started in the 1980s (Cunningham 2001). Gamma et al. (1995) gave a major contribution by their book on design patterns (a total of 23) which are organized into three **main categories** (listed here with names and intents).

Creational Patterns:

1. Abstract Factory: Provide an interface for creating families of related or dependent objects without specifying their concrete classes.

2. Builder: Separate the construction of a complex object from its representation so that the same construction process can create different representations.

3. Factory Method: Define an interface for creating an object, but let subclasses decide which class to instantiate. Factory Method lets a class defer instantiation to subclasses.

4. Prototype: Specify the kinds of objects to create using a prototypical instance, and create new objects by copying this prototype.

5. Singleton: Ensure a class only has one instance, and provide a global point of access to it.

Structural Patterns:

6. Adapter: Convert the interface of a class into another interface clients expect. Adapter lets classes work together that couldn't otherwise because of incompatible interfaces.

7. Bridge: Decouple an abstraction from its implementation so that the two can vary independently.

8. Composite: Compose objects into tree structures to represent part-whole hierarchies. Composite lets clients treat individual objects and compositions of objects uniformly.

9. Decorator: Attach additional responsibilities to an object dynamically. Decorators provide a flexible alternative to sub-classing for extending functionality.

10. Facade: Provide a unified interface to a set of interfaces in a subsystem. Facade defines a higher-level interface that makes the subsystem easier to use.

11. Flyweight: Use sharing to support large numbers of fine-grained objects efficiently.

12. Proxy: Provide a surrogate or placeholder for another object to control access to it.

Behavioral Patterns:

13. Chain of Responsibility: Avoid coupling the sender of a request to its receiver by giving more than one object a chance to handle the request. Chain the receiving objects and pass the request along the chain until an object handles it.

14. Command: Encapsulate a request as an object, thereby letting you parameterize clients with different requests, queue or log requests, and support undoable operations.

15. Interpreter: Given a language, define a representation for its grammar along with an interpreter that uses the representation to interpret sentences in the language.

16. Iterator: Provide a way to access the elements of an aggregate object sequentially without exposing its underlying representation.

17. Mediator: Define an object that encapsulates how sets of objects interact. Mediator promotes loose coupling by keeping objects from referring to each other explicitly, and it lets you vary their interaction independently.

18. Memento: Without violating encapsulation, capture and externalize an object's internal state so that the object can be restored to this state later.

19. Observer: Define a one-to-many dependency between objects so that when an object changes state, all its dependents are notified and updated automatically.

20. State: Allow an object to alter its behavior when its internal state changes. The object will appear to change its class.

21. Strategy: Define a family of algorithms, encapsulate each one, and make them interchangeable. Strategy lets the algorithm vary independently from clients that use it.

22. Template Method: Define the skeleton of an algorithm in an operation, deferring some steps to subclasses. Template Method lets subclasses redefine certain steps of an algorithm without changing the algorithm's structure.

23. Visitor: Represent an operation to be performed on the elements of an object structure. Visitor lets you define a new operation without changing the classes of the elements on which it operates.

A design pattern systematically names, motivates, and explains a general design that addresses a recurring design problem in object-oriented systems. It describes the problem, the solution, when to apply the solution, and its consequences. It also gives implementation hints and examples. The solution is a general arrangement of objects and classes that solve the problem. The solution is customized and implemented to solve the problem in a particular context.

Creational design patterns describe techniques for instantiating objects (or groups of objects). It deals with issues related to the creation of objects (Deitel et al. 2002). **Structural design patterns** describe common ways to organize classes and objects in a system. Developers often find two problems with poor organization. The first is that classes are assigned too many responsibilities. Such classes may damage information hiding and violate encapsulation, because each class may have access to information that belongs in a separate class. The second problem is that classes can overlap responsibilities. Burdening a design with unnecessary classes wastes time for designers because they will spend hours trying to extend or modify classes that should not even exist in the system. Structural design patterns help developers avoid these problems.

Behavioral design patterns assign responsibilities to objects. These patterns also provide proven strategies to model how objects collaborate with one another and offer special behaviors appropriate for a wide variety of applications. The Observer pattern is a classic example of collaborations between objects and of assigning responsibilities to objects. For example, GUI components use these patterns to communicate with their listeners, which respond to user interactions. A listener observes state changes in a particular component by registering to handle that component's events. When the user interacts with the component, that component notifies its listeners (also known as its *observers)* that the component's state has changed (e.g., a button has been pressed). A more detailed description of each pattern can be found at Cunningham (2001).

Mature engineering disciplines make use of thousands of design patterns. For example, a mechanical engineer uses a two-step, keyed shaft as a design pattern. Inherent in the pattern are attributes (the diameters of the shaft, the dimensions of the keyway, etc.) and operations (e.g., shaft rotation, shaft connection). An electrical engineer uses an integrated circuit (an extremely complex design pattern) to solve a specific element of a new problem. All design patterns can be described by specifying four pieces of information (Gamma et al. 1995):

- The name of the pattern,
- The problem to which the pattern is generally applied,
- The characteristics of the design pattern, and
- The consequences of applying the design pattern.

The design pattern name is an abstraction that conveys significant meaning about its applicability and intent. The problem description indicates the environment and conditions that must exist to make the design pattern applicable. The pattern characteristics indicate the attributes of the design that may be adjusted to enable the pattern to accommodate a variety of problems. These attributes represent characteristics of the design that can be searched (e.g. via a database) so that an appropriate pattern can be found. Finally, the consequences associated with the use of a design pattern provide an indication of the ramifications of design decisions.

The best designers in any field have an amazing ability to see patterns that characterize a problem and corresponding patterns that can be combined to create a solution. Gamma et al. (1995) discuss this when they state: "You'll find recurring patterns of classes and communicating objects in many object-oriented systems. These patterns solve specific design problems and make object-oriented design more flexible, elegant, and ultimately reusable. They help designers reuse successful designs by basing new designs on prior experience. A designer who is familiar with such patterns can apply them immediately to design problems without having to rediscover them".

Throughout the modeling process a builder should look for every opportunity to reuse existing patterns and to create new ones if reuse cannot be achieved.

2.7.3 GRASP patterns

GRASP is an acronym that stands for **G**eneral **R**esponsibility **A**ssignment **S**oftware **P**atterns. The GRASP patterns describe fundamental principles of assigning responsibilities to objects, expressed as patterns. What follows is a list of recognized GRASP patterns together with a short description (Larman 1998).

1. Expert
Who, in the general case, is responsible?

Assign a responsibility to the information expert - the class that has the information necessary to fulfill the responsibility.

2. Creator
Who creates?
Assign class B the responsibility to create an instance of class A if one of the following is true:
1. B contains A,
2. B aggregates A,
3. B has the initializing data for A,
4. B records A, or
5. B closely uses A.

3. Controller
Who handles a system event?
Assign the responsibility for handling a system event message to a class representing one of these choices:
1. The business or overall organization (a façade controller).
2. The overall "system" (a façade controller).
3. An animated thing in the domain that would perform the work (a role controller).
4. An artificial class (Pure Fabrication) representing the use-case (a use-case controller).

4. Low Coupling (evaluative)
How to support low dependency and increased reuse?
Assign responsibilities so that coupling remains low.

5. High Cohesion (evaluative)
How to keep complexity manageable?
Assign responsibilities so that cohesion remains high.

6. Polymorphism
When and whose behavior varies by type?
When related alternatives or behaviors vary by type (class), assign responsibility for the behavior to the types for which the behavior varies using polymorph operations.

7. Pure Fabrication
Who and when is someone desperate and does not want to violate High Cohesion and Low Coupling?
Assign a highly cohesive set of responsibilities to an artificial class that does not represent anything in the problem domain in order to support high cohesion, low coupling, and reuse.

8. Indirection
Who wants to avoid direct coupling?
Assign the responsibility to an intermediate object to mediate between other components and services, so that they are not directly coupled.

9. Don't Talk To Strangers (Law of Demeter)
Who to avoid knowing about the structure of indirect objects?
Assign the responsibility to a client's direct object to collaborate with an indirect object so that the client does not need to know about the indirect object. Within a method, messages can only be sent to the following objects:
- The object itself (Me),
- A parameter of the method,
- An attribute of Me,
- An element of a collection which is an attribute of Me,
- An object created within the method.

2.7.4 How to use design and GRASP patterns

In an object-oriented system, different mechanisms like inheritance and composition can use design and GRASP patterns. Inheritance is a fundamental object-oriented concept and was described in detail in paragraph 1.3. Using inheritance, an existing pattern becomes a template for a new subclass. Attributes and operations that exist in the design pattern become part of the subclass.

Composition is a concept that leads to aggregate objects. That is, a problem may require objects that have complex functionality (in the extreme, a system accomplishes this). The complex object can be assembled by a selected set of patterns and composing the appropriate object (or subsystem). Each pattern is treated as a black box, and communication among patterns occurs only via well-defined interfaces.

Gamma et al. (1995) suggest that object composition should be favored over inheritance when both options exist. Rather than creating large and maybe unmanageable class hierarchies (the consequence of the overuse of inheritance), composition favors small class hierarchies and objects that remain focused on one objective. Composition uses existing patterns (reusable components) in unaltered form.

2.7.5 Architectural patterns

Design patterns allow developers to design specific parts of systems, such as abstracting object instantiations, aggregating classes into larger structures or assigning responsibilities to objects. **Architectural patterns**, on the other hand, provide developers with proven strategies for designing subsystems and specifying how they interact with each other (Deitel et al. 2002).

The **Model-View-Controller** architectural pattern, for example, separates application data (contained in the *model)* from graphical presentation components (the *view)* and input-processing logic (the *controller).* In the design for a simple text editor, the user inputs text from the keyboard and formats this text using the mouse. The program stores this text and format information into a series of data structures, then displays this information on screen for the user to read what has been inputted. The model, which contains the application data, might contain only the characters that make up the document. When a user provides some input, the controller modifies the model's data with the given input. When the model changes, it notifies the view of the change so the view can update its presentation with the changed data—e.g., the view might display characters using a particular font with a particular size.

The **Layers** architectural pattern divides functionality into separate sets of system responsibilities called layers. For example, **three-tier applications**, in which each tier contains a unique system component, is an example of the Layers architectural pattern. This type of application contains three components that assume a unique responsibility. The information tier (also called the "bottom tier") maintains data for the application, typically storing the data in a data file or database. The client tier (also called the "top tier") is the application's user interface, such as a standard Web browser. The middle acts as an intermediary between the information tier and the client tier by processing client-tier requests, reading data from and writing data to the data store. Chapter 4 and all following examples in this book apply this type of architectural pattern.

Using architectural patterns promotes **extensibility** when designing systems, because designers can modify a component without having to modify another. For example, a text editor that uses the Model-View-Controller architectural pattern is extensible; designers can modify the view that displays the document outline but would not have to modify the model, other views or controllers. A system designed with the Layers architectural patterns is also extensible: designers can modify the information tier to accommodate a particular database product, but they would not have to modify either the client tier or the middle tier extensively.

2.8 An evaluation of OO approaches

Now it is time to review and evaluate the different object-oriented methods introduced so far. In summary, every OO approach models a problem or system by encapsulating data, functionality, and behavior in classes/objects that provide external interfaces to other classes/objects. A message-passing convention connects different classes/objects with each other and defines the communication channels inside of the model. Generally, OO methods emphasize concepts, notations, and strategies for modeling a problem/system. The same universal notation is used for both analysis and design in order to obtain a smooth transition between them.

More and more visual development tools are available to ease the OO development process and help with the implementation. For example, the following Web addresses offer free evaluation copies of modeling tools for downloading: http://www.XY.com, XY ∈ {pragsoft, rational}.

2.8.1 How to decide ?

The evaluation process can be organized around two types of questions: **before** and **after** a decision about object-oriented methods.

2.8.1.1. Questions before a decision could be:

- How good and helpful are OO methods at all?
- Why should I use an OO method if the same result can be achieved by a traditional method (e.g. functional, data-driven, or structured) I am familiar with?
- Why should I invest my time and energy in an approach that does not last very long and is just a fashion?

Some answers to these questions were already given in the introduction of chapter 1 and need not be repeated here. What is the major benefit of object-oriented methods? Most people agree it is the reuse of OO components as building blocks of a unit construction system for designing models and applications, thus allowing productivity to increase. Of course, not every problem should be modeled by using object orientation, and whether this approach will survive in the future or not, nobody can say. Depending on the type of problem/system, the use of more traditional methods like functional or data-driven techniques may be more successful. These traditional techniques have developed into what is nowadays called "structured methods" (Pressman 1997).

In the author's opinion, a decision should be based on the following **guidelines**:
- For smaller, more linear problems/systems use functional, data-driven, or structured approaches.
- For larger, more complex problems/systems use object-oriented methods.
- For problems/systems in between a combination of functional, data-driven or structured approaches with OO methods may be the best way.

Thus, the decision to select or reject an object-oriented method is a very hard one and requires thorough investigation. Further, some **guidelines** are needed in the decision making process which are based on some type of criteria. In any scientific discipline decision **criteria** are strongly related to the factors and corresponding questions listed in Table 2.3.

Table 2.3. Factors and questions to decide for/against an OO method

Factor	Question
Suitability	How suitable is the method under consideration to model a problem and implement the solution?
Financial costs	What is the cost to benefit ratio of the method under consideration for modeling a problem and implementing the solution?
Speed	How long does it take to model a problem and implement the solution with the method under consideration?
User-friendliness	How user-friendly is the method under consideration to model a problem and implement the solution?
Cost, size, speed, precision, robustness, safety of the solution	How expensive, large, fast, precise, robust and safe is the implemented solution with the used method?
Future development	How stable is the obtained solution in the future?

To most of these questions there are no clear answers available due to the intrinsic nature of the questions and a lack of definite results from research fields like "software metrics" (Fenton & Pfleeger 1996). But one thing seems to be sure; many projects (some guesses say more than half) where software engineering is a major part fail because something went wrong on the way from start to finish that is somehow related to the questions in Table 2.3.

Very often economical and organizational reasons prevent the successful completion of an engineering project. Unfortunately, this fact is either not enough known and taught in the engineering community, or too often neglected in engineering practice. In the real world, there is generally a lack of in-depth communication and understanding between the engineering and the business people involved in a project. Other common reasons for failure are:
- No or poor risk management,
- Missing the requirements, and
- Incorrect estimation of what can be achieved.

Therefore, careful investigation, research, planning, and project management of all relevant factors including those ones dealing with human beings is a must in order to prevent disasters!

But still there is no answer to many questions, such as the one regarding future developments. On the other hand, for example, it is very likely that the trend towards **network computing** will grow much stronger in the future. One indicator is the growth rate of the Web since its beginnings in 1995. After performance and security issues of networks are more satisfactorily solved, more and more engineering applications will probably take advantage of this technology. For example, the purchase of expensive engineering software with a user license could be avoided if the same program could be installed on the owner's server machine and ac-

accessed through a network by a client in another geographical location. Then the client just reimburses for the elapsed time of usage. Such a configuration is very beneficial for both sides; for the owner the problem of software piracy would disappear, and the client saves money needed for buying, maintaining, and updating software. Thus, network computing without object orientation is unthinkable.

2.8.1.2. Questions after a decision for object-oriented methods could be:

- What is the easiest OO method to start with?
- What is the best OO method available today?
- What are the advantages and disadvantages of each OO method?

Regarding the differences, advantages, and disadvantages of each OO approach, one can say that they share common object-oriented concepts, but are different in their notation and approach. Comparing Coad's method with the projection-type techniques of Booch, OMT, and the UML, it is obvious that each component in Coad's method could be as well realized by a corresponding object/class as part of a projection-type model. For example, a data management component in Coad's method and a data management superclass in the OMT can be designed in the same way with similar attributes and services. What may differ is the notation and the class organization inside of the complete model. The following Table 2.4 shows a comparison of different nouns with the same meaning (the OMT notation consists of much more nouns and corresponding symbols not included in this list).

Table 2.4. Different notations in OO methods with the same meaning

Coad's method	OMT and UML
Class	Class
Object	Object
Attribute	Attribute
Service	Operation
Connection	Link / Association
Whole-to-part connection	Link / Association based on Aggregation
Generalization-specialization connection	Link / Association based on Generalization-specialization

Coad's notation is less complicated or poorer than the OMT notation, depending on what an observer may judge as more important. Further, Coad's method seems to be more direct and natural whereas the projection-type methods have more potential to concentrate on certain aspects of the problem/system. The author recommends following the **guidelines** below:

- Beginners in object orientation without any modeling experience should use Coad's method while applying the UML notation.

- Beginners in object orientation with modeling experience using functional, data-driven, or structured approaches should use Booch's method, the OMT or the UML while applying the UML notation.
- Practitioners of object orientation should consider another type of method because they may gain new insights from it or find something useful.

2.8.2 Disadvantages of OO approaches

The **disadvantages** of OO approaches can be briefly summarized as follows. In comparison to functional, data-driven, or structured approaches, object-oriented methods are generally more difficult to:
- Learn due to their more complex nature,
- Apply when modeling a problem/system because their viewpoint is on a higher level of abstraction, and
- Implement because OO programming languages are not easier to learn and use than procedural languages (Java may be an exception to this).

Further, object-oriented methods have (Monroe et al. 1997):
- Difficulties in handling very large problems/systems,
- Difficulties in specifying how related collections of objects interact, and
- Difficulties in packaging related collections of objects for reuse.

In some cases the encapsulation of data, functionality, and behavior in discrete classes/objects can be a restriction in the OO modeling process because the different model parts are more seen in isolation, concentrating on their internal structure and external interfaces. Although these interfaces can reflect the formal behavior of each part in a problem/system, the overall behavior of the complete system can be so complex that it is hardly possible to capture the system's wholeness by just object-oriented methods.

CHAPTER 3: Fundamentals of Object-Oriented Models

3.1 Introduction: How to start and finish OO models?

The goal of this chapter is to describe fundamentals, strategies and guidelines for building and finishing successfully object-oriented models for a given problem/system. For this, the major modeling actions are to find, define, and create classes, objects, connections, attributes and operations as they are common to every object-oriented method. All these parts are essential for every OO model. The current chapter is organized around **to-do lists**, whereby list items are generally provided in the order of their sequence. But sometimes list items overlap and thus, their sequence is not totally clear in some cases. Therefore, each list item starts with a circle "•" as symbol and not with a consecutive number.

The modeling actions mentioned before belong to the technical or engineering actions of a development process. Economical and organizational aspects are as crucial in this process as the engineering ones for its **successful completion**. Unfortunately, this fact is either not adequately known and taught in the engineering community, or too often neglected in engineering practice. That is why so many engineering projects fail. Or viewed from an educational point of view, engineers are not sufficiently trained in business and project management, and business people are generally not educated in an engineering field. Thus in the real world, there is generally a lack of in-depth communication and understanding between the engineering and the business camps. Because the book is mainly written for engineers, this chapter includes also information about organizational strategies and project management as far as the author views them important for a successful object-oriented development project. The success of a project is mainly depending on the following factors:

- **Maintainability**: An implemented model should be quickly changeable, adaptable, and extensible to include new features.
- **Performance and reliability**: Testing of an implemented model should guarantee its quality but with shorter production times, its efficiency, its improved speed, memory consumption, and other performance factors.

- **Productivity**: An implemented model should be developed as fast as possible to meet the changing market, to reduce costs, and to be better than the competitors.
- **Reusability**: Time and effort to develop and implement a model can be saved by reusing existing units and solutions.
- **Safety**: safety is especially important for embedded, safety-critical systems and consumer goods.

Objects were defined in chapter 1 as instances of classes, which are templates for encapsulating data, functionality, and behavior with external interfaces to other classes/objects. Before starting to build an object-oriented model, names have to be given to these different items in the model. There are some general **rules and guidelines for naming** classes, objects, attributes, and operations (Norman 1996):
- Objects must always belong to a class.
- The first letter of a class name should begin with a capital letter; the same is true for each additional word in the same name. Examples: *GraphicSystem* or *GridPoint*.
- All names of classes, attributes, and operations should be singular. Example: *GridPoint* instead of *GridPoints*.
- All names of classes, attributes, and operations should be meaningful. Examples: initializeSystem or drawFrame.
- The class symbol is partitioned into three parts: name, attributes, and operations.
- Attribute and operation names should start with a lower case letter; each additional word used in the same name should begin with a capital letter. Examples: viewAngle or centerCoordinates.

The development process of a model, also called "project" in this book, can be categorized into **five phases:** 1) analysis, 2) planning, 3) design, 4) implementation, and 5) maintenance. Apart from a definition each phase is at least described by a to-do list of actions representing the strategy one should follow to ensure a more successful project.

3.1.1 Analysis

Analysis can be defined as the study of a problem/system leading to documented statements, specifications, and requirements for representing data, functionality, and behavior of the considered problem/system. It comprises all actions to:
- Write down a statement describing the problem/system to be modeled,
- Define the scope, purposes, and behavior of the problem/system,
- Apply use-cases as a set of prototypical scenarios to discover, capture, and document specifications and requirements (a requirement is a user demand that a system should fulfill),
- Collect knowledge about the problem/system under consideration,

- Understand what the system is doing,
- Explain by using natural language descriptions why the system behaves like as seen from an observer's viewpoint,
- Set up a model framework of specifications including assumptions, definitions, rules, and natural laws the problem/system obeys to,
- Move from more incomplete and inconsistent natural language descriptions to more complete, consistent, and precise model specifications,
- Incorporate mathematical or other scientific techniques, models, and tools if necessary or beneficial to refine the model requirements and specifications, and
- Select an object-oriented method to represent the model requirements and specifications, especially by using its graphical notation.

This list of actions reflects the major **principles of analysis** every method (functional, data-driven, or object-oriented) is using (Pressman 1997):
- The information extracted from a problem/system should be understood and represented by some type of notation.
- The analysis process should move from essential information toward finer details, or from the whole to its parts.
- Models that depict data, functionality, and behavior should be developed.
- Models should be partitioned in a way that uncovers detail in a hierarchial fashion.

During the analysis phase, four **tangible products** (artifacts) are created, each of equal importance (Booch 1996), which are a:
- Textual and graphical document of the problem/system containing its requirements and specifications,
- Collection of use-cases/scenarios defining the behavior and functionality of the problem/system,
- System model, and
- First risk assessment.

3.1.2 Planning

Planning can be defined as the definition, estimation, allocation, and optimization of resources, cost, and schedule for realizing a project. In the planning phase, a developer has to estimate:
- What and how many resources (people, hardware, software) are needed,
- How much the project will cost (finances),
- When it will start and how long it will take (time schedule), and
- What risks are involved?

Ideally, the analysis requirements and specifications are sufficient to make reasonable estimates. But very often this is not possible because the system to be modeled is ruled by an inverse problem, for example, that can only be solved by a trial-and-error method. Naturally, in such projects the risks can be very high and perhaps not worth taking. As defined, the planning of an engineering project comprises of all actions to:

- Define, estimate, allocate, and optimize needed people, hardware, software, and other resources,
- Define, estimate, allocate, and optimize the financial costs,
- Determine the overall architecture and configuration of the system,
- Develop a strategy, material, and tools how to build the model,
- Create a detailed action list,
- Set up a realistic time schedule,
- Design a risk management plan,
- Specify a quality assurance plan including validation criteria, and
- Develop a security plan.

Many of these actions belong at the same time to what is generally called "project management". The overall aim of **project management** is to avoid project failure and guarantee a successful completion of the development process. Its major tasks are planning, measurement, estimation, and control of all relevant factors like money, people, time, and tools in order to achieve pre-defined goals of the project (Henderson-Sellers 1997). For example, the action list in combination with the time schedule allows identification of temporal constraints and milestones, which can be used to measure intermediate progress of the project. Further, good project management realizes:

- The knowns and the unknowns of the project environment,
- What must be done to eliminate the uncertainties and unknowns, and
- What must be done to ensure that milestones are technically and politically feasible?

If necessary, good project management is able to re-plan the project and make midcourse corrections as required, whereby the trust and motivation of all people involved is kept at a very high level. To deliver incremental results at set milestones is of crucial importance in the project. This is why many developers prefer a rapid prototyping approach in their design, whereby a small portion of the whole model (preferably the core system, e.g. a graphical user interface) is initially created and evaluated by testing. Then adding other incremental parts gradually refines the model.

3.1.3 Design

Design can be defined as the selection of an architecture and building of a corresponding model before its physical implementation based on the documented requirements and specifications of the analysis and the results of the planning phase. It contains all actions to:

- Use the principles of object-oriented decomposition and reassembling to subdivide the whole model into subsystems or components,
- Decide how to translate the analysis requirements and specifications into subsystems of an object-oriented model,
- Decide what parts of the model can be reused or adapted from other sources,
- Build and test prototypes for the subsystems using the selected object-oriented method, and
- Reassemble and test the whole model using the selected object-oriented method.

The first action in this list is taken care of by all object-oriented methods described so far. For example, the OMT is based on object, dynamic, and functional models; Coad's method uses different components for the problem domain, human interaction, data management, and system interaction.

At the beginning of design, existing architectures should be considered first and how they can be reused in the present project before starting to build a new architecture from zero. When creating architectures, the focus is normally on three things: interfaces, an intelligent distribution of responsibilities throughout the system, and the exploitation of patterns that make the system simpler. During the design phase, five **tangible products** (artifacts) are generated (Booch 1996), which are a/n:

- Executable architecture,
- Specification of important architectural patterns,
- Release plan,
- Test criteria, and
- Revised risk assessment.

An executable architecture:
- Exists as a real application that runs in some limited way,
- Is production quality code,
- Carries out some or all of the behavior of interesting scenarios chosen from the analysis phase,
- Touches upon most if not all the key architectural interfaces,
- Makes a number of explicit assumptions, yet is not so simple that it ignores reality, and
- Either constitutes a vertical slice that cuts through the complete system from top to bottom, or goes horizontal by capturing most of the interesting elements of the system model.

An executable architecture forces the designer to tackle the pragmatics of all important system features. If the dominant risk to the project's success involves the system's technology, a vertical slice is preferable. If the major risk is instead the logic of the system, a horizontal slice is recommended. An executable architecture is the most important product of design because it:

- Serves as a tangible manifestation of many strategic decisions,
- Provides a platform for validating many important assumptions of analysis and design, and
- Offers a stable intermediate form that can be more easily developed when further changes are made or other features added.

Generally, design activities can be categorized as:
- High-level (also called macro design), including system architecture, component design, and interface design, for example, or
- Low-level (also called micro design), which includes algorithm design and data structure design.

High-level design focuses on decomposition and assembling while mapping requirements and specifications to design elements, whereas low-level design normally deals with coding. During design, three aspects should be always kept in mind:
- Simplicity of the model,
- A balanced distribution of responsibilities throughout the system, and
- Stability of the model.

3.1.4 Implementation

Implementation can be defined as the construction, testing, and installation of the designed model, making it available for users. If the goal is to develop computer models, computer-aided software engineering (CASE) tools are becoming more important because they allow faster prototyping and are less expensive. When it comes to model implementation by computer programming, this book relies more on visual development tools as used in part II of this book.

3.1.5 Maintenance

Maintenance comprises of all actions to:
- Evaluate the implemented model,
- Correct encountered errors in the implemented model,
- Update the model to accommodate changes in its environment, and
- Enhance the model to improve its functionality and performance.

The time and effort spent for maintenance directly reflects the goodness of an implemented model. This can be measured by the following factors (Booch 1996):
- Complexity is typically measured in terms of function points.
- Size is generally expressed by number of classes and/or lines of code.
- Stability is normally measured by the change rate in complexity or size.

- Quality is best measured in terms of absolute numbers of errors, defect discovery rate, and defect density.

3.1.5 Models of system development

All five phases combined in a linear sequential order build together the so called "waterfall model" (WFM) of system development shown in Fig. 3.1.

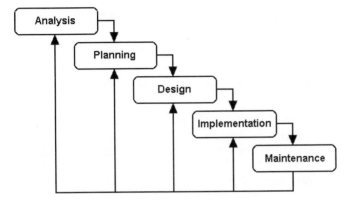

Fig. 3.1. Waterfall model of system development

The waterfall model is the oldest and most simple concept in engineering disciplines for developing a system. Only during the final phase of maintenance is information fed back to all previous development phases. Thus, this waterfall model has some major drawbacks, such as:

- Real projects hardly follow the sequential steps proposed by the WFM. Instead, iteration always occurs in every part of the development.
- All requirements and specifications are rarely known in the beginning of a modeling process. The WFM cannot incorporate this uncertainty.
- A working version of the model is very much delayed until the end of the project time span, which can be very dangerous if, for example, a serious fault is detected in the initial stages of the development.

To overcome these disadvantages of the WFM, a **spiral model** was designed by Boehm (1988) shown in Fig. 3.2. This model is a more realistic representation of what happens during the development process. It is basically the WFM extended into a spiral to allow iteration in the development.

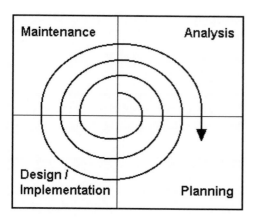

Fig. 3.2. Spiral model of system development

As can be seen in Fig. 3.2, the spiral model received its name from the spiral that symbolizes the iteration process from an initial prototype in the spiral's center to more improved model versions when moving outward the spiral. In this process, the model development is **an incremental, iterative, and controlled process** going continuously through four major phases drawn as four quadrants in Fig. 3.2:
1. Quadrant: analysis,
2. Quadrant: planning with an emphasis on risk management,
3. Quadrant: design and implementation, and
4. Quadrant: maintenance with an emphasis on evaluation.

During the first circuit around the spiral, initial development steps are taken with an initial model prototype to be evaluated at the end. After collecting suggestions for further model improvements, the next loop begins, and so on, until a successful model solution is found and implemented.

The spiral model can be further refined by what is generally called **round-trip engineering** or the round-trip model of system development, meaning the flexible process of continuous iteration back, forth and in between the different development phases as illustrated in the following figure.

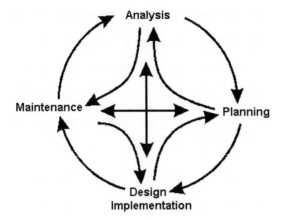

Fig. 3.3. Round-trip model of system development

Another way of categorizing the introduced phases of a development process is the following one (Pressman 1997):

1. Definition with the focus on **what** (corresponding to analysis and planning), for example:

- What is the system doing and what is the scope of the system?
- What are the necessary resources and costs for this project?
- What risks are involved in this project?

2. Development with the focus on **how** (corresponding to design and implementation), for example:

- How can the whole model be divided into subsystems or components?
- How can the analysis requirements and specifications be translated into subsystems or components of an object-oriented model?

3. Maintenance with the focus on **change**, which is the same as in the categorization introduced before. For example:

- What changes are required to correct, update, or enhance the implemented model?

Another model to describe the development phases is the framework of Zachman (1987) who introduced a simple but comprehensive formalism of organizing ideas and actions to create and deploy a model/system. A matrix as shown in Fig. 3.4 represents the Zachman framework.

	Data	Rules	Process	Network	People	Events
Analysis						
Planning						
Design						
Implementation						
Maintenance						
	What ?	Why ?	How ?	Where ?	Who ?	When ?

Fig. 3.4. The Zachman framework for organizing the development process of a model/system

Rows in the framework are identical with the various development phases whereas columns correspond to the distinct components of a model/system which can be identified by the different question words, such as what, why, how, where, who, and when. During the development process each cell in the matrix need to be addressed and the corresponding questions answered.

After having described strategies and guidelines for developing and implementing models based on five phases, major modeling actions dealing with classes, objects, connections, attributes and operations common to every object-oriented method will be considered in the next paragraphs.

3.2 Classes and objects

The identification and definition of classes and objects is the hardest part in the object-oriented development of a model (Booch 1994). Their identification and definition involves both discovery and invention. Discovery is especially needed during the phase of analysis to detect and extract all essential classes and objects. Invention is very important during the design phase when the behavior, interaction, and hierarchy of classes and objects are determined. Because of the various human factors involved in the design of a solution to a given problem/system, there is nothing like a "simple recipe" or a "perfect solution". As in any engineering discipline, design choices are a compromise between many competing factors.

Before looking at different approaches and strategies to identify classes and objects, it is helpful to keep the overall modeling process in mind, for example, by considering the question "How do you build models in general?" The answer incorporates at least the following actions:
- Observe, measure, make experiments, and collect data,
- Investigate and ask questions, especially those beginning with what, how, and why,
- Describe the behavior and functionality by e.g. mathematical tools,

- Assemble knowledge and understand the behavior and functionality using underlying rules, laws, and other cause-and-effect relations, and
- Develop models using the five different project phases introduced before: analysis, planning, design, implementation, and maintenance.

When identifying and defining classes and objects the following basic principles should be kept in mind valid for any successful engineering practice:
- Completeness: the problem/system should be decomposed in such a way that nothing important is left out.
- Formality: throughout the developing process rigorous and systematic methods should be applied.
- Simplicity: during the development process conceptual integrity and straightforwardness should be maintained.

When examining a problem/system from an object-oriented view point, one applies the four human rules of classification, generalization, and abstraction already described in paragraph 2.1. Historically, they have only been three general approaches to classification (Booch 1994):
- Classical categorization is based on the criteria that all entities in a category own one or more common properties.
- Conceptual clustering is a modern version of classical categorization where the selection criteria are not based on common properties, but on shared concepts among different entities.
- Prototype theory uses an entity of a category as a prototypical object, and the decision whether another object fits into the same category or not, depends on the degree of overlap between them. This theory emerged from the fact that some categories own neither clearly definable properties nor concepts, for example, games.

These approaches to classification build the foundation for different practices and rules of thumb briefly described below for identifying classes and objects. Further information can be found in the listed references, e.g. Booch (1994, 1996).

1. Classical analysis applies the principles of classical categorization to the requirements of a problem/system in order to define the involved classes and objects. Thus, classes/objects are categorized according to common properties. There are many options available for characterizing the attribute type of a class/object, for example, properties of:
- People involved,
- Physical things,
- Locations,
- Roles played,
- Events taking place,
- Interactions, or
- Concepts.

What properties need to be chosen for a specific problem/system depends very much on its natural characteristics.

2. Behavior analysis concentrates on the dynamic behavior of possible classes and objects following the approach of conceptual clustering. In this practice, classes are built on the criteria of similar behavior. For example, the *responsibility-driven approach* of Wirfs-Brock et al. (1990) can be directly used to form classes with objects that own the same responsibilities, such as clients and servers. This approach is directly related to the class-responsibility-collaboration (CRC) analysis, which identifies classes and objects where the behavior of candidate classes/objects is penciled on small, so called CRC cards.

3. Domain analysis consists of the identification, analysis, and specification of common requirements from a specific application domain, typically for reuse on multiple projects within that application domain, in terms of common classes, objects, subassemblies, and frameworks. The goal of domain analysis is straightforward. It is to find or create those classes that are broadly applicable, so that they may be reused. The role of a domain analyst is similar to the role of a master tool smith in a heavy manufacturing environment. The job of the tool smith is to design and build tools that may be used by many people doing similar, but not necessarily the same jobs. The role of the domain analyst is to design and build reusable components that may be used by many people working on similar, but not necessarily the same applications. Key inputs of the domain analysis are technical literature, existing applications, user surveys, expert advice and current/future requirements. Key outputs are class structures, reuse standards, functional models and domain descriptions. Domain analysis involves the following activities (Pressman 1997):

- Define the domain to be investigated.
- Categorize the items extracted from the domain.
- Collect a representative sample of applications in the domain.
- Analyze each application in the sample:
  - Identify candidate reusable objects,
  - Indicate the reasons that the object has been identified for reuse,
  - Define adaptations to the object that may also be reusable,
  - Estimate the percentage of applications in the domain that might make reuse of the object, and
  - Identify the objects by name, and use configuration management techniques to control them.
- Develop an analysis model for the objects.

In addition to these steps, the domain analyst should also create a set of reuse guidelines and develop an example that illustrates how the domain objects could be used to create a new application.

4. Scenario analysis or use-case analysis applies different scenarios within a problem/system domain to determine not only the classes and objects as parts of an object-oriented model but at the same time, the operations of each class/object are defined and how it collaborates with other classes/objects. By doing so, a scenario analysis brings together all the other approaches mentioned before.

5. Structured analysis can be used as a front end to object-oriented design because many engineers are trained and skilled in this type of analysis, and different development tools to speed up the work support it. But on the other hand, this analysis may be counter-productive because it does not lead automatically to an object-oriented model. In an object-oriented analysis the primary unit of decomposition is a class, not a function or an algorithm. Therefore, it needs very careful examination and in-depth understanding of both structured and object-oriented analysis to prevent the pitfalls of this approach. In general, structured analysis cannot be recommended for object-oriented analysis because of its different purpose.

6. Descriptive analysis is based on a written description of the problem/system under consideration and then underlining the nouns and verbs. The nouns represent possible objects and the verbs their possible operations. But because human language is not very accurate, and nouns and verbs can be phrased interchangeably, this analysis has the disadvantages of not being rigorous and not scaling well to larger problems or systems. Thus, it is more applicable to a small-sized problem/system.

All the introduced approaches for identifying classes and objects help the analyst to find a good enough initial solution. At this point it is important to remember that the analysis to identify classes and objects is the first incremental step in an iterative process of model development as shown in the spiral model (Fig. 3.2) or the round-trip model (Fig. 3.3).

3.3 Object-oriented analysis (OOA)

The objective of object-oriented analysis (OOA) is to develop a series of models that describe how an implemented model should work to satisfy a set of pre-defined requirements (Pressman 1997). OOA like conventional analysis methods builds a multipart analysis model to satisfy this objective. The analysis model depicts information, data, and behavior within the context of the different elements of the object model.

3.3.1 Convential versus OO approaches

Is object-oriented analysis really different from the structured analysis approach? Although debate continues, Fichman & Kemerer (1992) address the question head-on: "We conclude that the object-oriented analysis approach represents a radical change over process-oriented methodologies such as structured analysis but only an incremental change over data-oriented methodologies such as formation engineering. Process-oriented methodologies focus attention away from the inherent properties of objects during the modeling process and lead to a model of the problem domain that is orthogonal to the three essential principles of object orientation: encapsulation, classification of objects, and inheritance".

Stated simply, structured analysis takes a distinct input-process-output view of requirements. Data is considered separately from the processes that transform the data. System behavior, although important, tends to play a secondary role in structured analysis. The structured analysis approach makes heavy use of functional decomposition as can be seen in the partitioning of the data flow diagrams.

Fichman & Kemerer (1992) suggest 11 modeling dimensions that may be used to compare various conventional and object-oriented analysis methods:
1. Identification/classification of entities (in this context entity means either a data object regarding structered analysis or an object regarding OOA),
2. General to specific and whole to part entity relationships,
3. Other entity relationships,
4. Description of attributes of entities,
5. Large scale model partitioning,
6. States and transitions between states,
7. Detailed specification for functions,
8. Top-down decomposition,
9. End-to-end processing sequences,
10. Identification of exclusive operations, and
11. Entity communication (via messages or events).

Because many variations exist for structured analysis and OOA methods, it is difficult to develop a generalized comparison between the two methods. It can be stated, however, that modeling dimensions 8 and 9 are always present with system analysis.

3.3.2 An outline of different OO approaches

The popularity of object technologies has spawned dozens of OOA methods. Each of these introduces a process for the analysis of a product or system, a set of models that evolves out of the process, and a notation that enables the engineer to create each model in a consistent manner. In what follows, some of the more popular OOA methods are presented in outline form. The intent is to provide a snapshot of the OOA process. The UML is not included here because it is an evolution of the other OO methods.

3.3.2.1 Booch's method

The Booch's method (Booch 1994) encompasses both a micro and a macro development process. The micro level defines a set of analysis tasks that are re-applied for each step in the macro process. Hence, an evolutionary approach is maintained. The Booch's method is supported by a variety of automated tools. A brief outline of Booch's OOA micro development process follows:

- Identify classes and objects:
  - Propose candidate objects,
  - Conduct behavior analysis,
  - Identify relevant scenarios,
  - Define attributes and operations for each class.

- Identify the semantics of classes and objects:
  - Select scenarios and analyze,
  - Assign responsibility to achieve desired behavior,
  - Partition responsibilities to balance behavior,
  - Select an object and enumerate its roles and responsibilities,
  - Define operations to satisfy the responsibilities,
  - Look for collaborations among objects.

- Identify relationships among classes and objects
  - Define dependencies that exist between objects,
  - Describe the role of each participating object,
  - Validate by walking through scenarios.

- Conduct a series of refinements:
  - Produce appropriate diagrams for the work conducted above,
  - Define class hierarchies as appropriate,
  - Perform clustering based on class commonality.

- Implement classes and objects
  - Regarding OOA, this implies completion of the analysis model.

3.3.2.2 Object Modeling Technique (OMT)

Rambaugh et al. (1991) developed the Object Modeling Technique (OMT) for analysis, system design, and object-level design. The analysis activity creates three models: the object model (a representation of objects, classes, hierarchies, and relationships), the dynamic model (a representation of object and system behavior), and the functional model (a high-level data flow diagram like representation of information flow through the system). A brief outline of OMT's process follows:

- Develop a statement of scope for the problem.
- Build an object model:
    - Identify classes that are relevant for the problem,
    - Define attributes and associations,
    - Define object links,
    - Organize object classes using inheritance.

- Develop a dynamic model:
    - Prepare scenarios,
    - Define events and develop an event trace for each scenario,
    - Construct an event flow diagram,
    - Develop a state diagram,
    - Review behavior for consistency and completeness.

- Construct a functional model for the system:
    - Identify inputs and outputs,
    - Use data flow diagrams to represent flow transformations,
    - Develop process specifications for each function,
    - Specify constraints and optimization criteria.

3.3.2.3 Coad's method

The Coad's method as an extension of the Coad & Yourdon (1991) method is often viewed as one of the easiest OOA methods to learn. Modeling notation is relatively simple and guidelines for developing the analysis model are straight-forward. A brief outline of Coad's OOA process follows:

- Identify objects using "what to look for" criteria,
- Define a generalization-specification structure,
- Define a whole-part structure,
- Identify subjects (representations of subsystem components),
- Define attributes,
- Define operations.

3.3.2.4 Jacobsen method

The Jacobson (1992) method, also called "object-oriented software engineering" (OOSE) is a simplified version of the proprietary Objectory method, also developed by Jacobson. This method is differentiated from others by heavy emphasis on the use-case - a description or scenario that depicts how the user interacts with the product or system. A brief outline of Jacobson's OOA process follows:

- Identify the users of the system and their overall responsibilities.
- Build a requirements model:
  - Define the actors and their responsibilities,
  - Identify use-cases for each actor,
  - Prepare initial view of system objects and relationships,
  - Review model using use-cases as scenarios to determine validity.

- Build analysis model:
  - Identify interface objects using actor-interaction information,
  - Create structural views of interface objects,
  - Represent object behavior,
  - Isolate subsystems and models for each,
  - Review the model using use-cases as scenarios to determine validity.

3.3.2.5 Wirfs-Brock method

The Wirfs-Brock method (Wirfs-Brock et al. 1990) does not make a clear distinction between analysis and design tasks. Instead, a continuous process that begins with the assessment of a customer specification and ends with design is proposed. A brief outline of Wirfs-Brock's analysis-related tasks follows:

- Evaluate the user specifications,
- Use a grammatical parse to extract candidate classes from the specification,
- Group classes in an attempt to identify superclasses,
- Define responsibilities for each class,
- Assign responsibilities to each class,
- Identify relationships between classes,
- Define collaboration between classes based on responsibilities,
- Build hierarchical representations of classes to show inheritance relationships,
- Construct a collaboration graph for the system.

3.3.2.6 General steps for OOA

Although the terminology and process steps for each of these OOA methods differ, the overall OOA processes are really quite similar. To perform object-oriented analysis, a model builder should perform the following generic steps (Pressman 1997):

- Obtain user requirements with uses cases, scenarios and requirements,
- Select classes and objects using basic requirements as a guide,
- Identify attributes and operations for each system object,

- Define structures and hierarchies that organize classes,
- Build an object-relationship model,
- Build an object-behavior model, and
- Review the analysis model against use-cases and scenarios.

3.4 Class and object connections

After classes and objects are identified, they need to be connected to each other to ensure shared behavior as part of a complete model. Each class/object owns **three basic responsibilities** expressed by the following questions (Norman 1996):

1. Who a class/object knows?   => Connections,
2. What a class/object knows about itself? => Attributes, and
3. What a class/object does?   => Operations.

These questions refer to class/object connections, attributes, and operations, respectively which is the subject of this and the following two paragraphs.

Before considering class/object connections in more detail, it is helpful to keep in mind the following basic principles valid for designing any model:

- Abstraction: the problem/system should be represented in a simplified general from before adding more details that are not relevant for the model's functionality.
- Decomposition: larger problems/systems should be broken into smaller components which are again subdivided, and so on.
- Hiding: each component should be kept isolated and have only access to information that is required for its functionality. Hiding of classes and objects is achieved by the object-oriented concept of encapsulation.

Class/object connections are essential to create a model and make it work. They also help to make the model easier to comprehend, especially when the number of classes/objects reaches a larger size of more than fifty, for example. In such cases, it is necessary to connect classes and objects in an organized manner. One generally accepted way to do this is to apply the concept of patterns.

A **class/object pattern** is defined as a template for stereotypical class/object connections. Class/object patterns can be used over and over again by analogy as building blocks in order to increase the efficiency of object-oriented system development. They are commonly applied in various forms and styles, such as in art, architecture, industrial design, manufacturing, and many other disciplines. Two major class/object patterns were already described in chapter 2:

1. Generalization-Specialization, and
2. Whole-to-Part.

Object connections are another type of more specific class/object patterns introduced in paragraph 2.6 as notation symbols of Coad's method (Fig.2.27). Here is a list of some common object connections (Norman 1996):
1. Participant-Transaction,
2. Participant-Place,
3. Place-Transaction,
4. Transaction-Transaction line item,
5. Item-Transaction line item, and
6. Peer-Peer.

These object connections are especially important in information system modeling, and the following table shows some examples.

Table 3.1. Possible objects in an object connection pattern

Object	Example
Participant	Person, company, organization, ...
Transaction	Purchase, sale, payment, ...
Place	Store, office, bank, ...
Transaction line item	Sale line item, order line item, ...
Item	Product, service, information, ...
Peer	Objects within the same class

The labels "n" at both ends of the object connection in Fig.2.27 are the number of involved objects on both ends of the connection. They represent constraints on what other objects know about a specified object.

Class/object patterns are closely related to links and associations introduced as components of the object modeling technique in paragraph 2.4. There an **association** was defined as a group or a class of links with common structure, and a **link** as a connection between objects. Therefore, the words "association" and "class/object pattern" have more or less the same meaning and can be used interchangeable (compare Table 2.4). In the following, the two main class/object patterns, generalization-specialization and whole-to-part, will be explained in more detail.

A **generalization-specialization connection** is a hierarchical parent-child class pattern. Therefore, it serves to build a hierarchy of objects and classes that reflects the structure and architecture of the problem/system to be modeled. In UML notation, a line and a diamond symbol next to the general class represent it. Any combination and number of class and class-with-objects is technically allowed in this pattern with one exception. At the lowest level only class-with-objects symbols can be placed because classes without objects are not possible at this level.

The main characteristic of a hierarchical order is the concept of inheritance as described in paragraph 1.3. There **inheritance** was introduced as a key concept to express similarity of classes in a parent-child pattern to handle the complexity of a problem/system with the goal to simplify the desired model. All inheritance in this pattern is directed just in one way from the parent (general) class to the child (specialized) class. The child class inherits all attributes (data) and operations (functions) from the parent class. Therefore, there is no need in a class diagram to include the attributes and operations of a parent class again in a child class. Of course, the same is applicable to grandchildren classes, great-grandchildren, and so on, because the same parent-child pattern is true for each higher level in the class hierarchy. **Access specifiers** determine how to use class attributes and operations. They are: **private** (access only inside of the same class), **protected** (access only below the same branch of an inheritance tree), and **public** (everywhere else).

Normally, a specialized child class cannot be connected to any other classes than its own parent and children. But there are situations when such a rule maybe a hindrance like during the implementation stages of a model. For this, multiple inheritance was invented. **Multiple inheritance** is defined as the programming concept by which a child class inherits all attributes (data) and operations (functions) from more than one parent class. This concept has been discussed a long time because it is both very powerful and thus very dangerous at the same time. The author recommends avoiding multiple inheritance for reasons of safety and simplicity.

Another critical programming issue is the overriding of class attributes and operations. **Overriding** is defined as the programming practice of using the same names of attributes (data) and operations (functions) of a parent class in a child class, thus redefining and extending their meaning. Again, the author recommends avoiding the practice of overriding as much as possible for safety and simplicity reasons. At this point it is important to mention that the latest programming language Java[TM] introduced in 1995 by Sun Micro-systems avoids what some programmers call the "pitfalls" of a programming language like C++, which allows both multiple inheritance and operator overriding. In this way, Java is a much cleaner implementation language with fewer error-prone features, such as pointers or multiple inheritance.

The **whole-to-part connection** is as natural to humans as the generalization-specialization connection is. It allows building of connections between classes and objects according to a hierarchical whole-to-part class/object pattern. Such a pattern can be found in every physical item of some size, the smallest atomic particles excluded. In UML notation, a line and a triangle pointing in direction of the whole class represent the whole-to-part pattern. Any combination and number of classes and objects is possible in this pattern. The **main goals** of a hierarchy based on whole-to-part class/object patterns are the same as for parent-child patterns:

- Simplification of the problem/system under consideration, and
- Easier and more efficient modeling and implementation of the desired solution.

For the identification of whole-to-part class/object patterns three **basic categories** exist. They can be listed and described as follows:
1. Assembly and parts,
2. Container and contents, and
3. Group and members.

1. Assembly and parts is most likely the easiest category where a whole-to-part class/object pattern can be detected. As already mentioned before, every physical item of some size can be decomposed into smaller units. For example, a manufactured product (assembly) consists of different components (parts), such as a house is made up of walls, doors, windows, roof, etc.

2. Container and contents is a category very similar to assembly and parts, but with looser connections between single items and the difference that the container need not to be assembled for one overall purpose. Examples are offices (container) with tables, chairs, phones, file cabinets, etc. (contents), or factories with raw material, machinery, repair tools, etc. as contents. Generally, the purpose of a container is wider than the one of an assembly.

3. Group and members is another categorization where the other two types listed before do not work satisfactorily. This is especially true for groupings that are not directly visible by optical means. A professional organization (group) of workers (members) or a sports club (group) with sportsmen (members) are examples of this category. The next paragraph deals with attributes of classes and objects as integral parts of any object-oriented model.

3.5 Class and object attributes

Attributes (properties or data values) of classes and objects are essential to express their appearance and behavior. Without attributes classes and objects would lose their characteristics so that the development of a model would become impossible. Thus, attributes are integral and indispensable components of every object-oriented model. They also provide deeper insights into the identification of classes/objects and vice versa. Attributes add more details to classes and objects, which in turn reveals more about the overall class structure and architecture of a model.

Class and object attributes are normally manipulated by their own operations, the subject of the next paragraph. The recommendation and general rule is that the manipulation of attributes should only be performed by operations of the same class so that the object-oriented key concepts of encapsulation and information hiding are not violated. Or expressed in other words, each class or object has its own responsibility to manage and manipulate its own attributes. By doing so, a more

more simple and modular system development can be realized, benefiting more from the advantages of object orientation.

The identification of class and object attributes is an ongoing and iterative activity of the development process of a model. If an attribute like any other important element of a model was forgotten or wrongly modelled during the analysis and design phases of a project but later discovered, this mistake may be very difficult and costly to repair in later stages of the project. In this respect, attributes need a more careful examination than operations because attributes can have a much wider scope or area of influence within a class than operations.

There are different ways to investigate classes/objects and determine their corresponding attributes. A common practice is to think of yourself being placed into the position of a class or object, and then answer the following questions:
- How would "I" describe myself, if "I" were a maybe object in a problem   or system under consideration, for example, a computer interface?
- What do "I" know about myself, my own personal data, my appearance and properties?
- What information do "I" as a class or object need to know so that my responsibilities can be completed successfully?
- What data do "I" as a class or object need to remember over time so that   my responsibilities can be completed successfully? In other words,
- What states can "I" be in?

A **state** was defined in chapter 2 as the attribute values and links of an object, and an **event** as an action at a certain point in time that causes the state of an object to change. For instance, clicking the left mouse button over an icon causes a program to start executing and the icon to change its state from "deactivated" to "activated". The clicking of the left mouse button is the event, whereas the attributes "activated" and "deactivated" are the states of the icon.

In addition to the list above, many more questions can be asked to identify attributes of classes and objects, starting with "how", "when", or "why", for example. Once the attributes are found, their type has to be defined. There are at least three different **types of attributes**:
1. Single-value attributes,
2. Multi-value attributes, and
3. Dependent-value attributes.

1. A **single-value attribute** is defined as an attribute with only one value or state at any moment in time. It is probably the most frequently encountered attribute. Examples are the state of an icon as mentioned before, names, or scalar values in physics like density, temperature, pressure, etc.

2. A **multi-value attribute** is defined as an attribute with more than one value or state at any moment in time. It is an extension of the single-value attribute.

Again, there are many examples in physics for multi-value attributes, such as two- or three-dimensional tensors like 3D coordinates.

3. A dependent-value attribute is defined as an attribute with one or more values or states at any moment in time that depend on the values or states of other attributes. This type of attribute includes single-value and multi-value attributes, adding on dependencies on other attributes. Certain attributes of objects put constraints or limits on other ones, for example, the density and temperature of a physical body are related to each other.

3.6 Class and object operations

Operations of classes and objects, also called services, methods or procedures (functions), are the actions that a class or object has to perform to fulfill its intended purposes. Without operations classes and objects would become completely useless. Operations are initiated by or are the response to an event that takes place at a point in time. An event can be a user input or a request, for example. Such an event is often related to a change of state of the object affected by this event. During this process, messages are passed to and back from this object to invoke the operation and report back its successful completion or failure, respectively. In chapter 2, a message was defined as a request passed from a sender (a class/object) to a receiver (another class/object) that should provide a operation to satisfy the request. Thus, messages are tools by which classes/objects communicate with each other to accomplish their goals.

Considering this description of operations, events, states, and messages, the following **questions** can be asked for the identification of operations:
- What events are taking place, and what are the corresponding operations?
- How do the states or data values of objects change, and how can these changes be mapped into operations?
- What messages are sent to and back from the objects, and what are the associated operations of this message passing?
- What are the details of identified operations?

The last question may help to discover other operations not detected by the other questions if, for example, an operation needs assistance from another operation in the same or another object to complete its job. Such support operations can be very important in an object-oriented model.

There are two general categories of operations: basic operations and specific operations. **Basic operations** are implicit and automatically exist for every class. Sometimes, they are viewed so basic that they are not even included in an operation list of a class for reasons of simplicity. Depending on hardware specifications, operating systems, and implementation languages, some of the basic operations have to be explicitly defined and called at implementation time. They can be categorized as listed below in three groups of two operations, each being the opposite of the other operation:

1.
- Create object operation, also called "constructor",
- Delete object operation, also called "destructor",

2.
- Set object operation,
- Get object operation,

3.
- Add connection operation, and
- Remove connection operation.

The first group of basic operations creates and deletes object instances of a class. On calling the constructor operation, the considered object is defined and starts to exist physically by allocating computer memory and a memory address to it. The destructor operation removes an object physically from the computer memory so that there is no way to restore or recover the deleted object, unless a recycle bin is incorporated in the designed system as an extra software feature.

Once an object is created, any attributes of the object are zero or blank. For setting and getting object attributes the second group of basic operations is required. To pass data values to the attributes of an object or to update them, the set object operation is invoked. The set and get operations are the opposite of each other. The get operation retrieves the present attributes from an object whenever necessary, for example, when an attribute value is needed for a computation.

The third group of basic operations connects and disconnects objects from one another. They are used to couple or uncouple class/object connection patterns as described in paragraph 3.3. An object has to be disconnected first before it can be deleted properly.

Specific operations are not common to any form of classes, but unique to the problem/system and class they belong to. For example, a GUI class may own some special graphics operations. Basically, there is no restriction for designing specific operations. Any type and size is possible, meaning that a created operation should completely fulfill its given purpose. But it should be kept in mind that the key concepts of object orientation like encapsulation or information hiding are not violated. If a specific operation needs a support operation from the same or another class to complete its task, this operation can be invoked by sending out a corresponding message to it. In such a case, the requesting operation is called "sender operation", and the support operation receiving the request is considered the "receiver operation". After finishing its action(s), the receiver class sends the result back to the sender operation, which is then able to continue its work.

CHAPTER 4: Dynamic Matrix Processor in Visual Basic

4.1 Introduction to matrix processing

This and the following chapters have the goal to demonstrate how the theory of objects and patterns described in part I of this book can be used to create space-related applications in a simple manner. Of course, the selection can only be limited, but at least the most important topics are introduced and illustrated with up-to-date examples, such as graphical user interfaces, numerical computations, dynamic matrix processing, 2D and 3D graphics, databases, Java applets and parallel computing.

Another goal of all chapters in part II is to give readers an easy practical start into the subject. This is the reason why Visual Basic (VB) was chosen as programming language, at least in the beginning. VB is also an efficient tool to write code in a simple way for the first applications considered here. Further, VB allows incorporating objects and patterns in the software creation process (McMonnies 2001). This aspect of VB has been very often neglected in the past, but over the years VB has developed into a powerful alternative to other programming languages enriched by other technologies like ActiveX, Microsoft's distributed object and component model (Deitel et al. 1999).

VB does not try to emulate any OO programming language. Instead, it was initially produced to help developers to write programs for the Windows operating system (McMonnies 2001). In later versions it was decided to make the object-oriented core of VB more accessible to programmers. Thus, VB became a major exponent of object-oriented programming (OOP) for Microsoft Windows. Along with this development a **paradigm shift** took place
- from a static, text-oriented prompt environment
- to a dynamic, graphics-oriented window environment.

In such a window environment all graphical elements like forms, buttons or text boxes are objects of classes encapsulating attributes and operations. During run-time of such a window program objects are idle waiting to get into action by some type of event, either a user input (e.g. a mouse click) or another object. This is the

reason why writing programs for a window environment is also called "**event-driven programming**", consisting of the following steps (Schneider 1998):
• Decide how the windows that the user sees will look.
• Determine which events the objects on the windows should recognize.
• Write the operations ("procedures" in VB terminology) for those events.

Further, VB allows **dynamic memory management** for arrays of any type, which is very important for numerical computations. A static memory management of Fortran or other procedural programming languages has caused a lot of limitations in the past. With VB, memory of arrays can be flexible allocated, de-allocated and deleted during run-time. The program example in this chapter is based on this dynamic memory functionality of VB. "Dynamic matrix" means here more than one array of variable size.

All applications in part II use the **layers architectural pattern** introduced in chapter 2 in form of a three-tier architecture containing the three following layers as shown in the **use-case diagram** below (Fig. 4.1).
1. Top tier: a graphical user interfaces (GUI),
2. Middle tier: a processor, and
3. Bottom tier: a data manager.

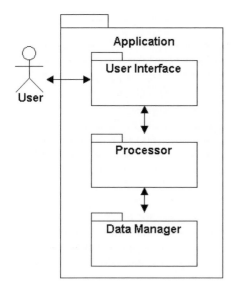

Fig. 4.1. Use-case diagram of a three-tier application

In this diagram, all tires are incorporated into packages, which are defined as containers of modeling elements in the UML notation. The different layers communicate with each other, and the top tier represents the interface that allows a user to communicate with the application. A possible realization of this architecture is shown in the following **class diagram** (Fig. 4.2), where all three tiers represent the top-level classes.

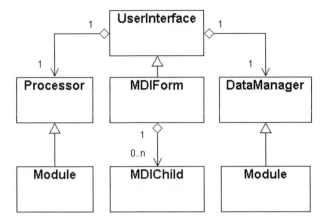

Fig. 4.2. Top-level class diagram of a three-tier application

Class *UserInterface* is derived from class *MDIForm* by a generalization-specialization association whereby classes *Processor* and *DataManager* have a whole-to-part association with *UserInterface*. Both classes *Processor* and *Data-Manager* are derived from class *Module*. The *MDIForm* (multiple document interface form) class allows creating a main window that acts as the backbone of the application and is the parent for *MDIChild* windows used to display the contents of one or more files, for example. *MDIChild* has a generalization-specialization association with class *Form*, which belongs to the Microsoft Foundation Class (MFC) library provided with the installation of VB. An *MDIForm* object consists generally of a menu bar, a tool bar, a status bar and a child window area.

When working through the different applications in the following chapters, you will realize that object orientation and patterns are just a **different way** to organize computer programs. The code itself consists mainly of assignments, loops and "If-Then-Else" clauses that remain largely the same for any type of programming method. The following text does **not(!)** include and explain every line of code in the program examples. Rather it concentrates on the main data and flow control shown in sequence diagrams in order to prevent information overload and thus confusion.

It is strongly recommended to try out and work through the following **program examples** no matter what your knowledge or experience of programming and/or VB is. Creating models and programs is, above all else, a practical subject. No book or learning aid will teach you to model and program unless you put the printed theory into practice (McMonnies 2001). Modeling and programming are crafts, such as cooking or painting. You have to practice it in order to gain enough knowledge and experience to become competent in it. Without practice, you may understand all the theory, but you will not be able to build and implement your own model!

All example programs incl. executable files and source code can be **downloaded** at http://de.geocities.com/bsttc2/book/SMOP.zip

If the executable files don't run properly due to a missing VB library file, you need to download and install the following **VB modules** (version 5) at:
http://de.geocities.com/bsttc3/dload/vb500a.zip
http://de.geocities.com/bsttc3/dload/vb500b.zip

The latter file "vb500b.zip" contains all VB database components. The source code examples can be compiled and run with VB version 5 or later.

4.2 Object-oriented analysis and design

The dynamic matrix processor considered here should fulfill the following **requirements**. It should allow to:
1. Open input files each containing a quadratic matrix **A** of variable size,
2. Compute the inverse matrix of $A = A^{-1}$, if **A** is a regular matrix,
3. Check the result with $A * A^{-1} = A^{-1} * A = I$, whereby **A** multiplied on either side with its inverse should produce an identity matrix, and
4. Show the result in the user interface and print them in an output file.

A use-case diagram for these requirements is given in Fig. 4.3.

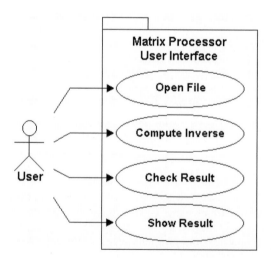

Fig. 4.3. Use-case diagram for a dynamic matrix processor

The class diagram of the dynamic matrix processor is based on the class diagram of the three-tier application (Fig. 4.2) plus some more classes to fulfill the given requirements. These classes are:

- class *CommonDialog* for selecting input and output files,
- class *Collection* for managing child windows, and
- class *MyFile* for file attributes and operations.

All three classes have a whole-to-part association with class *UserInterface* (*MyFile* through *MDIForm*). Classes *CommonDialog* and *Collection* are provided through the MFC library, whereas class *MyFile* is self-written. This is the reason why its name starts with "*My*". The class diagram of this application is shown below (from now on only the UML notation is used).

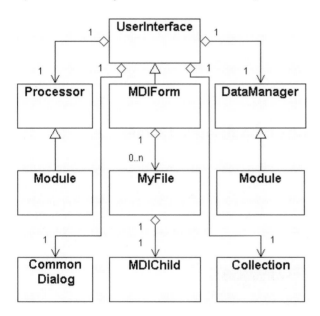

Fig. 4.4. Class diagram for a dynamic matrix processor

The scenario for the use-case diagram in Fig. 4.3 is shown in the following sequence diagram (Fig. 4.5).

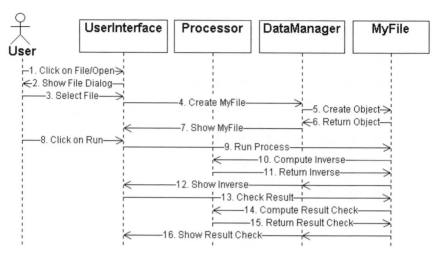

Fig. 4.5. Sequence diagram for a dynamic matrix processor

4.3 Implementation of the designed model

The coded architecture of the dynamic matrix processor can be easily visualized with the VB project explorer (McMonnies 2001) as shown in the figure below. This window is automatically displayed after opening the application's project file named "_Program01.vbp" in folder "Program01". All example programs incl. executable files and source code can be **downloaded** at http://de.geocities.com/ bsttc2/book/SMOP.zip

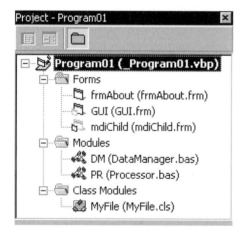

Fig. 4.6. VB project explorer window for the dynamic matrix processor

As can be seen in this figure, the project "Program01" in file named "_Program01.vbp" consists of three folders for Forms, Modules and Class Modules, each holding different classes with their corresponding files (file names in brackets). The mapping from the application's original class diagram (Fig. 4.5) to the coded diagram (Fig. 4.6) was done as follows.

Table 4.1. Original and coded class diagram for the matrix processor

Original (Fig. 4.5)	Coded (Fig. 4.6)	File name
UserInterface	GUI	GUI.frm
Processor	PR	Processor.bas
DataManager	DM	DataManager.bas
MyFile	MyFile	MyFile.cls
MDIChild	MdiChild	mdiChild.frm

The other class shown in Fig. 4.6 not listed in Table 4.1 is *frmAbout* in file "frmAbout.frm", which is used to display information about the application after clicking on "Help/About this program" in the program's menu bar.

Classes *Processor* and *DataManager* were implemented as VB "Modules" but not as "Class Modules" in "Program01" (Fig. 4.6), because their attributes and operations should be available with a global scope to all other objects all the time when the program is running. For smaller applications considered here, this approach is much more beneficial. If programmed as VB "Class Modules", they would be only accessible after corresponding objects were created and called.

How "Program01" works and how it was coded, can be more easily described by going through the application's sequence diagram (Fig. 4.5) and explaining step-by-step the corresponding program code. The start of "Program01" begins with the code shown in Table 4.2.

Table 4.2. Starting code in file "Processor.bas" in folder "Program01"

```
    ...
01  Option Explicit
02  Public fGUI As GUI                ' Declaration of public object fGUI of class
03  GUI Public ChildList As Collection ' Declaration of public object ChildList
04  RootDir As String                 '                        of class Collection
05
06  Sub Main()
07      Set fGUI = New GUI            ' Create fGUI and show it
08      fGUI.Show
09      Set ChildList = New Collection   ' Create ChildList
10      RootDir = CurDir
11  End Sub
    ...
```

The VB key words "Option Explicit" in line 01 of Table 4.2 tells the VB compiler to refuse all variables that have not been explicitly declared before being used. This enforces an important programming rule – that of declaring all variables. Lines 02-04 declare public objects "fGUI", "ChildList" and "RootDir" of classes *GUI*, *Collection* and *String* respectively. The VB procedure "Main()" between lines 06-11 is executed when "Program01" starts. Lines 07-10 initialize all these objects and display the main window via the public user interface object "fGUI".

Steps 1-7 in the application's sequence diagram (Fig. 4.5) for opening a file are implemented by the code shown in the following table.

Table 4.3. Code for "File/Open" in file "GUI.frm" in folder "Program01"

```
     ...
01 | Private Sub mnuFileOpen_Click()
02 | With FileDialog
03 |    .Filter = "Data Files (*.dat;*.txt)|*.dat;*.txt|Image Files (*.bmp;*.gif; _
04 |      *.jpg;*.wmf;*.emf)|*.bmp;*.gif;*.jpg;*.wmf;*.emf|All Files (*.*)|*.*;"
05 |    .DialogTitle = "Open an input file ..."
06 |    .MaxFileSize = 32000
07 |    .FileName = ""
08 |    .ShowOpen
09 |    If Len(.FileName) = 0 Then Exit Sub
10 |    If FileExists(.FileName) Then
11 |       Dim obj As Object                            ' Create new object
12 |       Set obj = CreateMyFile(.FileTitle, CurDir) '   and call creator operation
13 |    Else
14 |       MsgBox "File "' + .FileTitle + "' does not exist in folder " + CurDir + _
15 |                     "\ and cannot be opened.", vbInformation, fGUI.Caption
16 |    End If
17 | End With
18 | End Sub
     ...
```

After clicking on "File/Open" in the program's menu bar the operation in line 01 named "mnuFileOpen_Click()" is executed whereby an object named "FileDialog" of class *CommonDialog* is created and shown to the user (line 08). The filter of this object (lines 03-04) contains data text files, image files and all files with the corresponding file extensions. If the selected file exists (line 10), an object "obj" is created by calling an operation "CreateMyFile" in line 12, which takes two parameters: file title and file path (the current directory). Their combination gives the whole file name: FileName = FilePath + "\" + FileTitle. The code for operation "CreateMyFile" is shown in Table 4.4.

Table 4.4. "CreateMyFile" in file "DataManager.bas" in folder "Program01"

```
   ...
01 Public Function CreateMyFile(FTitle As String, FPath As String) As Object
02    Dim fi As MyFile                    ' Create new object
03    Set fi = New MyFile                 '      of class MyFile
04
05    Fi.FileID = Str(ChildList.Count + 1) ' Assign input file attributes
06       Fi.FileTitle = Ftitle
07       Fi.FilePath = Fpath
08       Fi.FileName = fi.FilePath + "\" + fi.FileTitle
09       If FileExists(fi.FileName) Then
10          fi.FileSize = FileLen(fi.FileName)
11       Else
12          fi.FileSize = 0
13       End If
14       fi.FileType = Right(fi.FileName, 3)
15       fi.FileStatus = True
16                                   ' Assign output file attributes
17       fi.OutFTitle = "Out " + fi.FileTitle
18       fi.OutFPath = fi.FilePath
19       fi.OutFName = fi.OutFPath + "\" + fi.OutFTitle
20                                   ' Assign child window attributes
21       fi.FileForm.FormID = fi.FileID
22       fi.FileForm.Caption = "File" + fi.FileID + ": " + fi.FileTitle
23       fi.FileForm.Text(0) = fi.FileTitle
24       fi.FileForm.Text(1) = fi.FilePath
25       fi.FileForm.Text(2) = Str(fi.FileSize)
26       fi.FileForm.Text(3) = fi.FileType
27       fi.FileForm.Text(4) = fi.OutFTitle
28       fi.FileForm.Text(4).SetFocus
29
30       ChildList.Add Item:=fi                  ' Add fi to ChildList
31
32       Call fi.ReadData                        ' Read fi's data
33
34 Set CreateMyFile = fi
35 End Function
   ...
```

Lines 02-03 declare and create an object named "fi" of class *MyFile* including a child window, which is an *mdiChild* object named "FileForm", used to display the opened file and its data (step 7 in Fig. 4.5). The assignment of "fi" attributes happens in lines 05-28 depending on the given file parameters "FTitle" and "FPath" passed on to this function in line 01. These attributes of "fi" contain properties of the selected input file (lines 05-15), an output file (lines 17-19) as well as attributes of "FileForm" (lines 21-28). Each new "fi" receives a unique ID named "FileID" in line 05 given by the number of items in object "ChildList" of class *Collection* incremented by +1. This "FileID" is a *String* object (the VB procedure "Str" converts a number to a string) identical to the ID of the child window object "FileForm" named "FormID" (line 21). "FileID" is also displayed in the title bar of each child window (line 22 and Fig. 4.7).

After completing the attribute assignment in Table 4.4, the new object "fi" is added to the "ChildList" in line 30 and its data is read in line 32. Finally, object "fi" is set equal to object "CreateMyFile" (line 33) and passed back to the calling object in line 12 of Table 4.3.

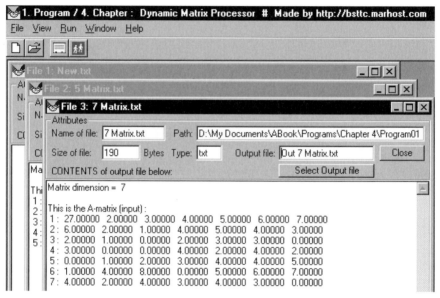

Fig. 4.7. Child windows after step 7 in the sequence diagram (Fig. 4.5)

As mentioned before, class *Collection* is provided through the Microsoft Foundation Class (MFC) library, which allows an easy and simple management of items in a list, here objects of class *MyFile*. A *Collection* object provides a convenient way to refer to a related group of items as a single object. The items, or members, in a collection need only to be related by the fact that they exist in the collection. Members of a collection don't have to share the same data type. Once a collection is created, members can be added using the "Add" method and removed using the "Remove" method. Specific members can be returned from the collection using the "Item" method, while the entire collection can be iterated using the "For Each...Next" statement. When closing a child window and thus deleting a *MyFile* object, for example, the code in the following table removes this object from the "ChildList" and re-indexes the remaining items in this list.

Table 4.5. Unloading a child window in file "mdiChild.frm" in folder "Program01"

```
    ...
01  Private Sub Form_Unload(Cancel As Integer)
02      Dim obj, i As Long
03      For Each obj In ChildList                    ' Remove child from list
04          If Val(FormID) = Val(obj.FileID) Then ChildList.Remove Val(FormID)
05      Next obj
06      i = 1
07      For Each obj In ChildList                    ' Re-index children in list
08          obj.FileID = Str(i)
09          obj.FileForm.FormID = Str(i)
10          obj.FileForm.Caption = "File" + obj.FileID + ": " + obj.FileTitle
11          i = i + 1
12      Next obj
13  End Sub
    ...
```

For finding the correct item in a list of *MyFile* objects in line 04, the fact is used that the attribute "FormID" of an *mdiChild* object named "FileForm" is equal to the "FileID" of a *MyFile* object according to lines 05 and 21 in Table 4.4 (the VB procedure "Val" converts a string to a number). These attributes are re-indexed in lines 08-09 and then updated in the title bar of each child window in line 10.

The data input for a *MyFile* object is realized by calling an operation "ReadData" in line 32 in Table 4.4. The code of this public routine of class *MyFile* is given in the table below.

Table 4.6. Code for "ReadData" in file "MyFile.cls" in folder "Program01"

```
    ...
01      Public Sub ReadData()
02      If FileExists(FileName) Then
03
04      If LCase(FileType) = "txt" Or LCase(FileType) = "dat" Then
05          Dim i As Integer, j As Integer, AnyNumber As Single
06          i = 0
07          Open FileName For Input As #1      ' Open file for input
08              Do While Not EOF(1)            ' Loop until end of file
09              Input #1, AnyNumber
10              i = i + 1
11              Loop
12          Close #1                           ' Close file
13          Adim = Sqr(i)                      ' Compute matrix dimension
14
```

```
15        ReDim A(Adim, Adim)
16        Open FileName For Input As #1 ' Input data of matrix A
17           For i = 1 To Adim
18              For j = 1 To Adim
19                 Input #1, A(i, j)
20              Next j
21           Next i
22        Close #1              ' Print A in form and output file
23        Call PrintMat(A, Adim, "This is the A-matrix (input) :", _
24                   FileForm.Text1, OutFName, 0)
25     End If
26
27     If LCase(FileType) = "bmp" Or LCase(FileType) = "gif" _
28     Or LCase(FileType) = "jpg" Or LCase(FileType) = "wmf" _
29     Or LCase(FileType) = "emf" Then
30        FileForm.Picture1.Picture = LoadPicture(FileName)
31        FileForm.Height = FileForm.Picture1.Height + 1620
32        FileForm.Picture1.Visible = True
33        FileForm.Text1.Visible = False
34     End If
35
36     Else
37        Adim = 0
38     End If
39     End Sub
          ...
```

After checking the existence of file "FileName" of a *MyFile* object, "If" clauses are used to distinguish different file types in lines 04 and 27-29. Here two types can be read: data text and image files. Of course, this routine can be easily extended to read other file types like databases or executable files, which will be done in applications considered later.

If the input file is a data file with extensions "dat" or "txt", lines 05-23 are executed to determine the dimension of the quadratic matrix **A**, which is the square root of the total amount of numbers in the input file (line 13), and read its data depending on this variable dimension (lines 16-22). The dynamic memory management of matrix **A** is done in line 15 with the VB "ReDim" procedure. Then the dimension and the matrix itself are printed in an object named "Text1" of class *TextBox* inside of the child window "FileForm" and the output file by calling the operation "MatPrint" (line 23-24 and Fig. 4.7), which is a public operation of the *DataManager* module. If the input file is an image file with extensions "bmp", "gif", "jpg", "wmf" or "emf", the image is loaded into an object named "Picture1" of class *PictureBox* belonging to object "FileForm" (line 30). Both classes *TextBox* and *PictureBox* are also provided through the MFC class library. Then object

"Picture1" is resized (line 31), "Picture1" made visible and object "Text1" made invisible (lines 32-33).

A detailed description of the implementation of steps 1-7 in the application's sequence diagram (Fig. 4.5) was included here to illustrate how the programming was done and how the code is working. At this point it is also recommended to look again at the sequence diagram in Fig. 4.5 in order to comprehend more clearly how the application is functioning till to this point. The remaining steps in the sequence diagram (Fig. 4.5) represent two loops to compute the inverse of matrix **A** (1. loop: steps 9-12) and to check the result (2. loop: steps 13-16), after the user clicked on "Run" in the applications's menu bar (step 9) to start the processing. This executes the code in the table below.

Table 4.7. Code for "File/Run" in file "GUI.frm" in folder "Program01"

```
       ...
01  Private Sub mnuRun_Click()
02      Dim obj
03      For Each obj In ChildList    ' Find child in list
04          If Val(Screen.ActiveForm.FormID) = Val(obj.FileID) Then
05              obj.RunProcess        ' 1. Loop: steps 9-12
06              obj.CheckResult       ' 2. Loop: steps 13-16
07          End If
08      Next obj
09  End Sub
       ...
```

After finding the active child window using again the fact that the attribute "FormID" of an *mdiChild* object is equal to the "FileID" of a *MyFile* object according to lines 05 and 21 in Table 4.4, the inverse of **A** and its control is done by two public operations (lines 05-06) of a *MyFile* object shown in the following table. These two operations "RunProcess" and "CheckResult" do the major data processing in this application.

Table 4.8. Public operations "RunProcess" and "CheckResult" in file "My-File.cls" in folder "Program01"

```
        ...
01  Public Sub RunProcess()
02  If FileExists(FileName) Then
03      If LCase(FileType) = "dat" Or LCase(FileType) = "txt" Then
04          ReDim Ainv(Adim, Adim)
05          Dim i As Integer, j As Integer
06          For i = 1 To Adim
07              For j = 1 To Adim
08                  Ainv(i, j) = A(i, j)
09              Next j
```

```
10            Next i
11
12        Call BuildInverse(Ainv, Adim)
13        If Adim > 0 Then ' Check for regular matrix
14            Call PrintMat(Ainv, Adim, "This is the inverse of A (output) :", _
15                        FileForm.Text1, OutFName, 1)
16        Else
17            Dim nl As String
18            nl = Chr(13) & Chr(10)
19            FileForm.Text1.Text = FileForm.Text1.Text + nl + nl + _
20                        "This matrix is not regular!"
21        End If
22
23      End If
24   End If
25   End Sub
26
27   Public Sub CheckResult()
28   If FileExists(FileName) Then
29      If LCase(FileType) = "dat" Or LCase(FileType) = "txt" Then
30          Dim C() As Single
31          ReDim C(Adim, Adim) ' CONTROL A*A-1
32          Call MultiplyMat(C, A, Ainv, Adim, Adim, Adim)
33          Call PrintMat(C, Adim, "This is the result of A*A-1 (control) :", _
34                      FileForm.Text1, OutFName, 1)
35                          ' CONTROL A-1*A
36          Call MultiplyMat(C, Ainv, A, Adim, Adim, Adim)
37          Call PrintMat(C, Adim, "This is the result of A-1*A (control) :", _
38                      FileForm.Text1, OutFName, 1)
39      End If
40   End If
41   End Sub
        ...
```

In the beginning, both operations check the existence of file "FileName" of a *MyFile* object and its data type, which has to be "dat" or "txt". Then operation "RunProcess" initializes the inverse matrix of **A** (lines 04-10) and computes the result in line 12. If **A** is regular, routine "BuildInverse" returns the original value of "Adim", otherwise a zero. If the returned value of "Adim" is greater than zero (line 13), the inverse matrix is printed in the child window and output file (lines 14-15). Otherwise an error message is shown (17-20). Because "Adim" is equal to zero in this case, the "CheckResult" operation does not compute and show any matrices.

The "CheckResult" operation starts by declaring and dimensioning a third matrix **C** (lines 30-31) for keeping the results of $A * A^{-1} = A^{-1} * A = I$, whereby **A** multiplied on either side with its inverse should produce an identity matrix. The results are printed using "PrintMat", a public operation of the *DataManager* module. The other two routines listed in Table 4.7 are "BuildInverse" and "Multiply-Mat", both public operations of the *Processor* module (see files "DataManager.bas" and "Processor.bas" in folder "Program01" for further details).

Another routine very much used in this application is "FileExists" (Table 4.3, 4.4. 4.6 and 4.8) which takes a *String* object "FileName" as parameter as shown in the table below. If the file "FileName" exists, the return value of this routine is True, otherwise False. The check is performed by using the VB procedure "Dir$" in line 03.

Table 4.9. Code for routine "FileExists" in file "DataManager.bas" in folder "Program01"

```
    ...
01  Public Function FileExists(FileName As String) As Boolean
02      Dim retval As String
03      Retval = Dir$(FileName)
04      If retval = "" Then
05          FileExists = False
06      Else
07          FileExists = True
08      End If
09  End Function
    ...
```

4.4 Patterns used in this application

As already described in the introduction 4.1, the **layers architectural pattern** containing a user interface, a processor and a data manager is the fundamental three-tier architecture of all applications considered in part II of this book (Fig. 4.1 and 4.2).

Design patterns used in this application are:

Creational Patterns:

2. Builder: Separate the construction of a complex object from its representation so that the same construction process can create different representations.

3. Factory Method: Define an interface for creating an object, but let subclasses decide which class to instantiate. Factory Method lets a class defer instantiation to subclasses.

4. Prototype: Specify the kinds of objects to create using a prototypical instance, and create new objects by copying this prototype.

Behavioral Patterns:

16. Iterator: Provide a way to access the elements of an aggregate object sequentially without exposing its underlying representation.

19. Observer: Define a one-to-many dependency between objects so that when an object changes state, all its dependents are notified and updated automatically.

The **builder, factory method and prototype patterns** are used in this application when opening files (Table 4.3) and calling the "CreateMyFile" operation (Table 4.4), which is a factory method. Complex and different *MyFile* objects are created in the same way. Only the "ReadData" operation decides on which subclasses to instantiate (Table 4.6). These subclasses are *TextBox* for data text files and *PictureBox* for image files, for example. In this way, class *MyFile* represents a prototype that can be copied and easily extended to open other file formats. This will be demonstrated when the first database program is introduced in chapter 8.

In this application considered here, the **iterator pattern** is being used with the *Collection* object "ChildList" in order to manage and access child windows as part of *MyFile* objects (Table 4.5 and 4.7). The items, or members, in a collection need only to be related by the fact that they exist in the collection. Members of a collection don't have to share the same data type. Once a collection is created, members can be added using the "Add" method and removed using the "Remove" method. Specific members can be returned from the collection using the "Item" method, while the entire collection can be iterated using the "For Each...Next" statement.

The **observer pattern** is automatically included in the one-to-many dependency between a MDI parent and its child windows through a VB "WindowList" attribute of a *Menu* object, for instance. This attribute allows a *Menu* object to maintain a list of MDI child windows in an *MDIForm* object (Fig. 4.2). If set to True, the *Menu* object maintains a list of open child windows and displays a check mark next to the active window. The list for Fig. 4.7, for example, is shown below.

Fig. 4.8. Child window list of Fig. 4.7 using the VB "WindowList" attribute

Here the *Menu* object "Window" was chosen to display the child window list. A user can click a window name to activate any child window in this list. Automatically the child window, that was active before, is notified, set to inactive and its display is updated.

In this application, all GRASP patterns are used exept the last three ones listed in paragraph 2.7.3: pure fabrication, indirection and law of demeter.

4.5 Testing the application

The dynamic matrix processsor to compute the inverse of a matrix **A** with a variable size received from different data text files, was tested in various ways in order to fulfill the requirements given in the beginning of paragraph 4.2. These tests can be categorized according to the application's three-tier architecture as follows:
- Testing the user interface,
- Testing the processor, and
- Testing the data manager.

Tests of the user interface concentrated on the child window management using the *Collection* class. When adding new items to a *Collection* object like in line 30 in Table 4.4, for example, it was found that using a key by writing

30 │ ChildList.Add Item:=fi, Key:=fi.FileID

was not beneficial because this key causes an error when not the last child window in a list was closed and then a new one added.

Testing the processor and the data manager was straightforward. If a matrix is not regular, the following message is displayed to inform the user.

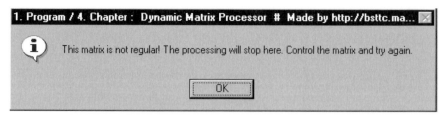

Fig. 4.9. Information message for a singular matrix

This message is produced by operation "BuildInverse" in module *Processor* (see file "Processor.bas" in folder "Program01" for further details), if a zero diagonal element is encountered during the Gaussian elimination process (Koch 1988). If a matrix is regular, the following result is shown in the child window and the selected output file (Fig. 4.10).

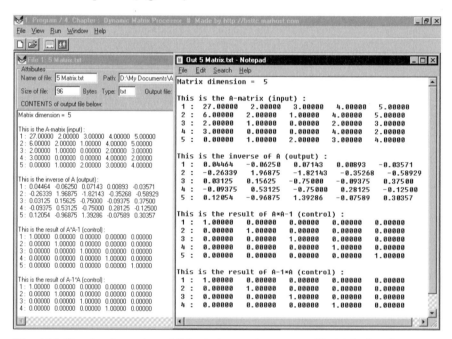

Fig. 4.10. Results shown in a child window and an output file (right-hand)

Input and output files can be directly viewed by clicking on "View/ Input File" or "View/Output File" in the menu bar so that the following code inside of routines "ViewInputFile" and "ViewOutputFile" is called:

Table 4.10 Code for viewing directly a data text file in file "MyFile.cls" in folder "Program01"

	...
01	Dim o
02	o = Shell(RootDir + "\Notepad.exe " + FileName, vbNormalFocus)
	...

The VB "Shell" procedure runs an executable program (here "Notepad.exe") inside of the application's root directory and returns a value representing the program's task ID if successful, otherwise it returns zero. When the file name is added to the parameter string of "Shell" like in line 02, the executable program directly opens this file.

CHAPTER 5: 2D Dynamic Data Plotter in Visual Basic

5.1 Introduction to 2D graphics in Visual Basic

The application considered in this chapter follows very much the previous one in terms of the general architecture and how the designed model works in principle. The **main difference** is the incorporation of graphical concepts and components to be able to visualize and print any contents of a 2D data file through a 2D plot of this data. "Plotter" means here viewer and printer. To make it more interesting, a best-fitting line and a Fourier analysis supplement the application.

With the advent of digital computers **data visualization** has become an essential part of daily life (e.g. to display the development of consumer prices, or when viewing stock market charts). It transforms data from a digital format into an analog format so that it can be more easily checked, evaluated and interpreted when viewed through human eyes. Thus, data visualization is a very useful application and the reason why it was selected here. It also illustrates how a new program can be created in a simple manner by reusing and extending an already existing one (here "Program01").

All applications in part II of this book use a **file as the main building block** (*MyFile* as the base class) to create applications containing a parent / multiple child object system as the underlying design structure (Fig. 4.2). Here a file can be of the following type:
- Data text (ASCII) files,
- Graphics or image files,
- Database files, or
- Executable files with "exe" as file extension.

It will be demonstrated how these different file types can be processed by building on **just one base class (*MyFile*)** and retain the general design and implementation of a variety of window programs. The main reason for this is the seamless application of design patterns (mainly the builder, factory method and prototype patterns) already used in "Program01" and described in paragraph 4.4. For example, the programs in this and the last chapter can simultaneously open and read data text and graphics files. By extending this approach to database and executable files, a simple and powerful design method is at hand to build various applications and satisfy many different needs at the same time.

Visual Basic provides all the tools needed **to generate and print graphical images**. It supplies routines for drawing text, lines, rectangles, circles and other graphical shapes. The VB "Point" and "PSet" procedures allow specifying the exact value of every pixel in an image, giving absolute control over the result (Stephens 2000).

The three most important items for drawing in VB are objects of classes *Form*, *PictureBox* and *Printer*. These objects support a wide variety of drawing procedures that can be used to produce graphics. They provide almost exactly the same graphic operations, so most of the things you can do with one you can do with the others.

To begin drawing in VB, the coordinate system (Fig. 5.1) needs to be explained used by all drawing objects. By default, the upper left corner of a drawing object has the coordinates (x, y) = (0, 0). A 2D coordinate pair is composed of a horizontal X coordinate and a vertical Y coordinate. The X coordinate is the horizontal distance moving right from the upper-left corner. The Y coordinate is the vertical distance moving down from the upper-left corner.

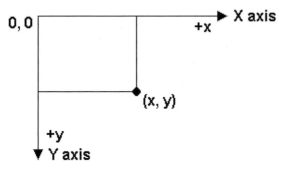

Fig. 5.1. Coordinate system for *Form*, *PictureBox* and *Printer* drawing objects

The scale of a drawing object's coordinate system can be selected by setting the object's "ScaleMode" attribute to one of the values in Table 5.1. The many different scales available in VB support flexibility in the graphics programming. The units in Table 5.1 are specified in theoretical inches. A **theoretical inch** is an inch as it is theoretically displayed on a printer. Its length on your actual monitor or printer depends on your hardware and is probably not exactly one real inch (Stephens 2000).

Table 5.1. Possible VB "ScaleMode" attribute settings

ScaleMode	Meaning
vbInches	Measurements in theoretical inches
vbMillimeters	About 25.4 millimeters per theoretical inch
vbCentimeters	About 2.54 centimeters per theoretical inch
vbTwips	1440 twips per theoretical inch
vbPixels	The smallest unit on a computer screen (a single dot). A 14" screen with a resolution of 800 x 600 pixels, for example, has one pixel of 25.2 twips (= 1440/800*14).
vbPoints	72 points per theoretical inch. One point has 20 twips.
vbCharacters	120 twips horizontally and 240 twips vertically per character

By default, VB is using "vbTwips" as "ScaleMode". This coordinate system gives sufficient precision for many applications, but sometimes another system is more convenient. For example, the VB procedure "Scale" given by

```
01 | Object.Scale (Xmin, Ymax) – (Xmax, Ymin)
```

scales a drawing object to (Xmax-Xmin, Ymax-Ymin), reverses the direction of the Y coordinate axis, and sets the coordinate center to (Xmin, Ymin), which is the lower-left corner of the drawing object. Thus, the "Scale" procedure allows creating a coordinate system on a drawing object as normally wanted.

Many advanced graphics techniques use Windows Application Programming Interface (**API**) functions. These functions always measure coordinates in pixels. If you use API functions in a drawing area, you may want to set the drawing area's "ScaleMode" to "vbPixels". Then you will not need to convert back and forth between pixels for the API functions and whatever other scale you use for VB drawing procedures. It is also usually easier to work with pixels when a program needs exact control over every pixel (Stephens 2000).

There are several ways for **sending output to a *Printer* object**. The VB "Print-Form" procedure is the easiest one, but it also produces the poorest quality and can be very slow. By using graphical VB procedures like "Line" or "Circle", graphics can be directly send to a printer. This needs more programming code, but it produces better output, and a printer generally prints faster. Another advantage of the latter method is to include a a print preview feature to let users see what a printout will look like before printing it. This will be also included in the program of this chapter.

VB allows only one single *Printer* object to exist. This is an example for the **Singleton** creational design pattern described in paragraph 2.7.2. All *Printer* objects in a Windows system are listed in the VB "Printers" collection. The code below, for example, displays all printers in a message box (lines 04-07) and sets the *Printer* object to the first printer in the "Printers" collection (line 09).

```
01 │ Dim ptr As Printer
02 │ Dim str As String
03 │
04 │ For Each ptr In Printers
05 │   str = str & ptr.DeviceName & Chr(13)
06 │ Next ptr
07 │ MsgBox str
08 │
09 │ Set Printer = Printers.Item(0)
```

The message box in line 07 shows normally more printers than expected because objects including fax software and graphical subsystems like from Adope Systems Inc. can appear as printers to the Windows system.

A VB *CommonDialog* object makes it very easy to select a *Printer* object. For example, the following table displays the " mnuFilePrintSetup_ Click()" operation executed after a click on "File/Print Setup" in the program's menu bar.

Table 5.2. Code for the print setup dialog in file "GUI.frm" in folder "Program02"

```
        ...
01 │ Private Sub mnuFilePrintSetup_Click()
02 │     With FileDialog
03 │         .ShowPrinter
04 │     End With
05 │ End Sub
        ...
```

Lines 02-04 create and display a *CommonDialog* object named "FileDialog" so that a user can select a printer and its attributes, such as its page size, orientation (portrait or landscape), paper source, print quality or number of copies.

A *Printer* object has several procedures needed to control printing. Some of the most important are "**NewPage**", "**EndDoc**", and "**KillDoc**". "NewPage" starts a new page and resets "CurrentX" and "CurrentY" printer coordinates to the upper-left corner of the new page. "EndDoc" finishes the current print job and sends it to the physical printer. Until "EndDoc" is executed, an output to the printer is buffered and does not appear. The "KillDoc" operation is used to delete a waiting print job.

A **print preview** is a very useful feature for every program that allows printing. A simple way to produce a print preview is to draw all the print procedures into a *PictureBox* object and then display it to the user together with a response message box. If the user response is positive, the same is send then to a *Printer* object. The problem with this method is that *PictureBox* and *Printer* objects use different attributes, whereby their width and height are the most important parameters. To minimize the programming effort, the best is to write just one printing routine that takes as parameters the named items: a drawing object (a *PictureBox* or *Printer* object) as well as its width and height. Exactly this was done with the program in this chapter. The printing operation is called "PlotToPrinter" and works as follows. In the beginning, a print preview is shown with a *PictureBox* object as parameter of "PlotToPrinter". If the user wants a printout, the same operation is called again, but with the *Printer* object as parameter. After the operation finishes, the program executes the *Printer* object's "EndDoc" procedure to send the printout to the physical printer.

A very good alternative to a printout via the program's print routine is to save the drawing as a **printout image file** and use it from there as wanted, for example, inserting it into a word processor document. This functionality is very advantageous when writing a document where the image is a natural part of the whole text. In general, this method is even more required and therefore included in the program of this chapter. The corresponding operation called "mnuFile-SaveAs_Click()" is executed after clicking on "File/Save Plot As" in the program's menu bar (see paragraph 5.3 for further details).

As already mentioned before, the 2D dynamic data plotter in this chapter is based on the same three-tier architecture of the previous program "Program01" in chapter 4, containing the three following layers as shown in the **use-case diagram** in Fig. 4.1:
1. Top tier: a graphical user interfaces (GUI),
2. Middle tier: a processor, and
3. Bottom tier: a data manager.

The corresponding top-level **class diagram** for this architecture is shown in Fig. 4.2. Thus, there is no need to show both figures 4.1 and 4.2 again because there are also true for the new program called "Program02" in this chapter.

5.2 Object-oriented analysis and design

The 2D dynamic data plotter considered here should fulfill the following **requirements**. It should allow to:
1. Open 2D input files of variable size,
2. Display a 2D data plot with a best-fitting line for each input file,
3. Compute a Fourier analysis for each input file, and
4. Show the result of the Fourier analysis in the user interface and print it in an output file.

The class diagram of the 2D dynamic data plotter is based on the class diagram of the dynamic matrix processor in chapter 4 (Fig. 4.4) plus some other classes added to class *MyFile* (Fig. 5.2). These classes are:

- class *MyPoint* to create 2D data points and their coordinates,
- class *PointList* to manage *MyPoint* objects, and
- class *mdiPrint* for displaying print previews.

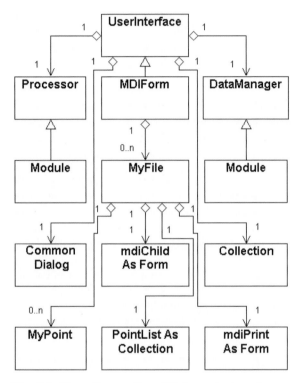

Fig. 5.2. Class diagram for a 2D dynamic data plotter

All three new classes have a whole-to-part association with class *MyFile*. *MyPoint* is a self-written class with an ID and X,Y coordinates as attributes, whereas *PointList* and *mdiPrint* have a generalization-specialization association with classes *Collection* and *Form* respectively, which belong both to the MFC library. The resulting class diagram is shown in Fig. 5.2.

A sequence diagram for the given requirements of this application is given in Fig. 5.3.

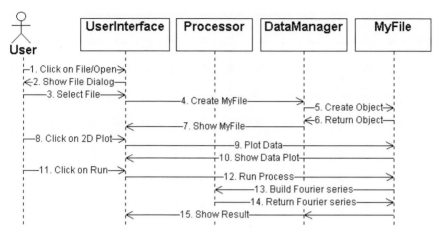

Fig. 5.3. Sequence diagram for a 2D dynamic data plotter

This diagram is very similar to the sequence diagram of the dynamic matrix processor (Fig. 4.5) in the last chapter. One major difference is that after selecting an input file two user clicks are required to produce the final result (a Fourier analysis) in Fig. 5.3 in comparison to just one click in Fig. 4.5 to compute and check the inverse of a regular matrix. Another difference is the incorporation of *SSTab*, *CheckBox* and *ComboBox* objects in the *mdiChild* object, all by a whole-to-part association and provided also through the MFC class library. These classes were not included in the class diagram (Fig. 5.2) to preserve simplicity and readability. The *SSTab* object placed on a child window allows switching back and forth between the digital file output of a *MyFile* object (Tab:"Contents") and its analog graphical output (Tab: "2D Plot") by clicking on the corresponding tab (Fig. 5.4).

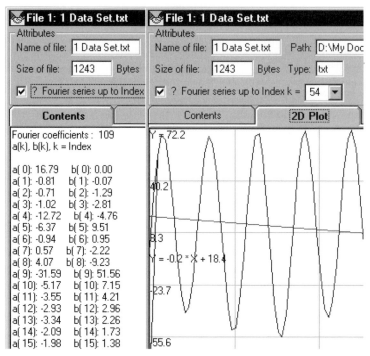

Fig. 5.4. An *mdiChild* object including *SSTab*, *CheckBox* and *ComboBox* objects as part of a 2D dynamic data plotter

In this figure, the *CheckBox* object is on and shows the caption: "? Fourier series up to Index k =", whereas the *ComboBox* object contains all possible values for k. This application computes a Fourier series with the following formulas (function values f(t) should have an equal distance dt):

$$a_0 = \frac{1}{n}\sum_{t=1}^{n} f(t) \ ,$$

$$a_k = \frac{1}{n}\sum_{t=1}^{n} f(t)\cos\left(\frac{\pi}{n}kt\right), \qquad b_k = \frac{1}{n}\sum_{t=1}^{n} f(t)\sin\left(\frac{\pi}{n}kt\right) \ , \qquad (5.1)$$

$$f(t) = \frac{a_0 + a_m\cos(\pi t)}{2} + \sum_{k=1}^{m-1}\left[a_k\cos\left(\frac{\pi}{n}kt\right) + b_k\sin\left(\frac{\pi}{n}kt\right)\right].$$

where a_k and b_k are the so called Fourier coefficients with
n = number of data points = 2 * m + 1,
m = order of Fourier series = (n − 1) / 2, and
k = index of Fourier series $\in$ {0, 1, 2, …, m}.

5.3 Implementation of the designed model

The coded architecture of the 2D dynamic data plotter can be easily visualized with the VB project explorer (McMonnies 2001) as shown in the figure below. This window is automatically displayed after opening the application's project file named "_Program02.vbp" in folder "Program02". All example programs including executable files and source code can be **downloaded** at http://de.geocities.com/bsttc2/book/SMOP.zip

Fig. 5.5. VB project explorer window for the 2D dynamic data plotter

As can be seen in this figure, "Program02" is an extension of "Program01" (Fig. 4.6) with new classes and forms *MyPoint* in file "MyPoint.cls" and *mdiPrint* in file "mdiPrint.frm". These classes have been added to the project according to the class diagram in Fig. 5.2.

How "Program02" works and how it was coded, can be more easily described by going through the application's sequence diagram (Fig. 5.3) and explaining step-by-step the corresponding program code. Because the beginning of "Program02" is almost identical to "Program01" until step 7 (Fig. 4.5), a description of this part is not necessary where the contents is identical, which is the case for Tables 4.2, 4.3 and 4.4. The first difference occurs in operation "ReadData" of a *MyFile* object with the following programming code (Table 5.3).

Table 5.3. Code for "ReadData" in file "MyFile.cls" in folder "Program02"

	...
01	Public Sub ReadData()
02	If FileExists(FileName) Then

```
03
04      If LCase(FileType) = "txt" Or LCase(FileType) = "dat" Then
05         Dim i As Integer, j As Integer, xp As Single, yp As Single
06         i = 0
07         Set PointList = New Collection ' Create PointList
08         Open FileName For Input As #1 ' Open file for input
09            Do While Not EOF(1)      ' Loop until end of file
10            i = i + 1
11            Input #1, xp, yp
12            Dim poi As MyPoint       ' Create new object
13            Set poi = New MyPoint   '  of class MyPoint
14            poi.ID = Str(i)
15            poi.X = xp
16            poi.Y = yp
17            PointList.Add Item:=poi   ' Add poi to PointList
18            Set poi = Nothing
19            Loop
20         Close #1      ' Close file
21         Max = i      ' Compute maximum number of data points
22
23         Call PrintList(PointList, Max, "This is the point list (input) :", _
24                         FileForm.Text1, OutFName, 0)
25         Call PlotData(FileForm.Picture1)
26      End If
27
28      If LCase(FileType) = "bmp" Or LCase(FileType) = "gif" _
29      Or LCase(FileType) = "jpg" Or LCase(FileType) = "wmf" _
30      Or LCase(FileType) = "emf" Then
31         FileForm.Picture1.Picture = LoadPicture(FileName)
32         FileForm.Tab1.Height = 350 + FileForm.Picture1.Height
33         FileForm.Height = 1600 + FileForm.Tab1.Height
34         FileForm.Tab1.Width = FileForm.Width - 135
35         FileForm.Picture1.Width = FileForm.Tab1.Width
36         FileForm.Tab1.Tab = 1
37      End If
38
39  Else
40         Max = 0
41         FileForm.Check1 = 0
42         FileForm.Tab1.Tab = 0
43  End If
44  End Sub
        ...
```

After checking the existence of file "FileName" of a *MyFile* object, "If" clauses are used to distinguish different file types in lines 04 and 28-30. Here two types can be read: data text and image files. If a data text file exists, a *Collection* object named "PointList" (a public member class of a *MyFile* object according to Fig. 5.2) is created in line 07 and filled with 2D data points read from file "FileName" (lines 08-20). For doing this, a *MyPoint* object named "poi" is created in lines 12-13, assigned an ID as well as X and Y coodinates (lines 14-16) and then added to object "PointList" in line 17. Then "PointList" is used to print data in the child window area of the user interface and in an output file (lines 23-24). Further, calling an operation "PlotData" of a MyFile object in line 25 creates a 2D data plot.

If the input file is an image file with extensions "bmp", "gif", "jpg", "wmf" or "emf", the image is loaded into a *PictureBox* object named "Picture1" belonging to the child window object "FileForm" (line 31). Then objects "Picture1" and the *SSTab* object called "Tab1" are resized (line 32-35) and the second tab of "Tab1" with an index equal to one is made visible (line 36).

If the input file "FileName" of a *MyFile* object does not exist at all, the total number of data points "Max" is set to zero (line 40), the *CheckBox* object of named "Check1" to off (line 41), and the *SSTab* object "Tab1" to zero (line 42), thus showing the "Contents" tab (Fig. 5.4).

The "PlotData" operation in line 25 of Table 5.3 has a code given in the following table. A plot result was already shown in Fig. 5.4.

Table 5.4. Code for "PlotData" in file "MyFile.cls" in folder "Program02"

```
       ...
01   Public Sub PlotData(obj As Object)
02   If FileExists(FileName) Then
03       If LCase(FileType) = "dat" Or LCase(FileType) = "txt" Then
04         If PointList.Count > 0 Then
05           obj.Cls   ' Clear drawing object
06
07         Dim i As Integer, Sum(4) As Single, xp As Single, yp As Single
08         Dim poi
09         Set poi = PointList.Item(1)          ' Start initialization
10           Par(1) = poi.X                     ' Min for X
11           Par(2) = poi.Y                     ' Min for Y
12           Par(3) = poi.X                     ' Max for X
13           Par(4) = poi.Y                     ' Max for Y
14
15         For i = 1 To 4
16           Sum(i) = 0                         ' Initialize Sum values
17         Next i
18         For Each poi In PointList
19           If Par(1) > poi.X Then Par(1) = poi.X
20           If Par(2) > poi.Y Then Par(2) = poi.Y
```

```
21      If Par(3) < poi.X Then Par(3) = poi.X
22      If Par(4) < poi.Y Then Par(4) = poi.Y
23      Sum(1) = Sum(1) + poi.X              ' Sum(1) for X
24      Sum(2) = Sum(2) + poi.Y              ' Sum(2) for Y
25      Sum(3) = Sum(3) + poi.X * poi.X         ' Sum(3) for X*X
26      Sum(4) = Sum(4) + poi.X * poi.Y         ' Sum(4) for X*Y
27    Next poi
28
29      obj.Scale (Par(1), Par(4))-(Par(3), Par(2))    ' Scale drawing object
30      For i = 1 To 6                          ' Draw X axis in 5-er distance
31        xp = Par(1) + (i - 1) * (Par(3) - Par(1)) / 5
32        obj.Line (xp, Par(2))-(xp, Par(4)), RGB(220, 220, 220)
33        obj.CurrentX = xp
34        obj.CurrentY = Par(2) + 200 * (Par(4) – Par(2)) / obj.Height
35        If i = 6 Then obj.CurrentX = obj.CurrentX _
36              - 140 * (Par(3) - Par(1)) / obj.Width _
37              - 70 * Len(Format(xp, "###0.0")) * (Par(3) - Par(1)) / obj.Width
38        If i = 1 Then
39          obj.Print "X = " + Format(xp, "###0.0")
40        Else
41          obj.Print Format(xp, "###0.0")
42        End If
43              ' Draw Y axis in 5-er distance
44        yp = Par(2) + (i - 1) * (Par(4) - Par(2)) / 5
45        obj.Line (Par(1), yp)-(Par(3), yp), RGB(220, 220, 220)
46        obj.CurrentX = Par(1)
47        obj.CurrentY = yp
48        If i = 1 Then obj.CurrentY = obj.CurrentY _
49                              + 400 * (Par(4) - Par(2)) / obj.Height
50        If i = 6 Then
51          Obj.Print "Y = " + Format(yp, "###0.0")
52        Else
53          Obj.Print Format(yp, "###0.0")
54        End If
55      Next i
56
57    Set poi = PointList.Item(1) ' Draw lines from point to point
58      xp = poi.X
59      yp = poi.Y
60    For Each poi In PointList
61      obj.Line (xp, yp)-(poi.X, poi.Y), RGB(0, 0, 0)
62      xp = poi.X
63      yp = poi.Y
64    Next poi
65                      ' y = ax + b, a:=Par(5), b:=Par(6)
66      Dim Si As String  ' Best-fitting line (in red)
```

```
67 | Par(5)= (Max*Sum(4) – Sum(1)*Sum(2)) / (Max*Sum(3) - Sum(1)* Sum(1))
68 | Par(6) = (Sum(2)*Sum(3)–Sum(1)*Sum(4))/(Max*Sum(3) - Sum(1)*Sum(1))
69 |       obj.Line (Par(1), Par(5) * Par(1) + Par(6))- _
70 |               (Par(3), Par(5) * Par(3) + Par(6)), RGB(255, 0, 0)
71 |       obj.CurrentX = Par(1)
72 |       obj.CurrentY = (Par(2) + Par(4)) / 2 + 100 * (Par(4) - Par(2)) / obj.Height
73 |       Si = "+"
74 |       If Par(6) < 0 Then Si = "-"
75 |       obj.Print "Y = " + Format(Par(5), "###0.0") + " * X " + Si _
76 |               + " " + Format(Abs(Par(6)), "###0.0")
77 |
78 |     End If
79 |   End If
80 | End If
81 | End Sub
   |   ...
```

The code in Table 5.4 for the "PlotData" operation of a *MyFile* object is structured as follows. After checking the existence of file "FileName", its extension "FileType" and the number of points in "PointList" (three "If" clauses starting in lines 02-04 and ending in lines 78-80), the transferred drawing object named "obj" (line 01) is cleared in line 05. Here, "obj" is a *PictureBox* object "Picture1" of an *mdiChild* object "FileForm" according to line 25 in Table 5.3.

The computational part of operation "PlotData" starts in line 07 with the initialization of all variables needed to scale "obj" with the smallest and largest coordinate values ("Par()" in lines 10-13) and their summation values for the best-fitting line ("Sum()" in lines 15-17). After computing the "Par()" and "Sum()" variables in lines 18-27, the drawing object "obj" is scaled in line 29 using the corresponding "Par()" values. Then a coordinate system is drawn and values of coordinate axis are written in 5-er steps (lines 30-55). This is followed by drawing a black line from data point to data point (lines 57-64) and completed by inserting the best-fitting line drawn in red color (lines 66-76).

The operation "PlotData" in Table 5.4 is also called after a click on the "2D Plot" tab of the *SSTab* object when the *CheckBox* object value is false (Fig. 5.4), which represents steps 8-10 in the sequence diagram in Fig. 5.3.

But if the *CheckBox* object value is true like shown in Fig. 5.4, a Fourier series can be computed, checked and results displayed in blue color (steps 11-15 in the sequence diagram in Fig. 5.3). There are two options available to do this. If the *CheckBox* object value is true, a user can click on:
• "File/Run" in the program's menu bar, or
• "2D Plot" tab of the *SSTab* object.

Further, one can click on the *CheckBox* object so that its value changes from false to true to get a Fourier analysis. Moreover, a click on the *ComboBox* object in Fig. 5.4 allows displaying the calculated Fourier series with a smaller index k in order to **test which Fourier coefficients are relevant** and which ones can be neglected.

If the *CheckBox* object value is false, only the "PlotData" operation in Table 5.4 is executed. The code for a click on "File/Run" is shown in the table below.

Table 5.5. Code for "File/Run" in file "GUI.frm" in folder "Program02"

```
    ...
01  Private Sub mnuRun_Click()
02      Dim obj
03      For Each obj In ChildList 'Find child in list
04          If Val(Screen.ActiveForm.FormID) = Val(obj.FileID) Then
05              Screen.MousePointer = vbHourglass
06              obj.FileForm.Tab1.Tab = 1
07              obj.RunProcess  ' Build and show Fourier series
08              Screen.MousePointer = vbDefault
09          End If
10      Next obj
11  End Sub
    ...
```

After finding the active child window using the fact that the attribute "FormID" of an *mdiChild* object is equal to the "FileID" of a *MyFile* object according to lines 05 and 21 in Table 4.4, the mouse pointer is changed to an hour glass (line 05) and the Fourier series is computed and shown in lines 06-07 using the "RunProcess" operation of a *MyFile* object (Table 5.6).

Table 5.6. Code for "RunProcess" in file "MyFile.cls" in folder Program02"

```
    ...
01  Public Sub RunProcess()
02  If FileExists(FileName) Then
03      If LCase(FileType) = "dat" Or LCase(FileType) = "txt" Then
04        If PointList.Count > 0 Then
05          Call BuildFourier(PointList, Max, Fa, Fb)
06          Call PrintResult(PointList, Max, "Fourier coefficients : ", _
07                      FileForm.Text1, OutFName, 0, Fa, Fb)
08          FileForm.Check1 = 1
09          FileForm.Combo1.Clear
10          Dim i As Integer, n As Integer
11          n = (Max - 1) / 2
12          For i = 0 To n
13              FileForm.Combo1.AddItem Str(i)
```

```
14        Next i
15        FileForm.Combo1.ListIndex = n
16        Call ShowResult(FileForm.Picture1)
17      End If
18    End If
19 End If
20 End Sub
   ...
```

After checking the existence of file "FileName", its extension "FileType" and the number of points in "PointList" (three "If" clauses starting in lines 02-04 and ending in lines 17-19), the Fourier series is computed in line 05 and printed in the child window area of the user interface and in an output file (lines 06-07). Then the *CheckBox* object named "Check1" is set to true (line 08) and the *ComboBox* object named "Combo1" is filled with the index numbers k of the Fourier series (lines 09-15). In line 16, this series is plotted in a *PictureBox* object "Picture1" of an *mdiChild* object "FileForm" by calling the *MyFile* operation "ShowResult".

As already mentioned in paragraph 5.1, there are two options in this application available to export or print the graphical output:
- Image file outputs by clicking on "File/Save Plot As", or
- Hardcopy printouts by clicking on "File/Print" in the program's menu bar.

The programming code of both options together with a description how they work is given below.

Table 5.7. Code for "File/Save Plot As" in file "GUI.frm" in folder "Program02"

```
   ...
01 Private Sub mnuFileSaveAs_Click()
02 If ChildList.Count > 0 Then
03      Dim obj
04      For Each obj In ChildList            ' Find child in list, then Exit loop
05          If Val(Screen.ActiveForm.FormID) = Val(obj.FileID) Then Exit For
06      Next obj
07      Dim oldFileName As String, FileName As String, FileTitle As String, _
08                      ext As String, i As Integer
09      oldFileName = obj.FileName
10      ext = Right(oldFileName, 3)
11      With FileDialog
12          .Filter = "Image Files (*.bmp)|*.bmp|All Files (*.*)|*.*;"
13          .DialogTitle = "Save Plot As ..."
14          .MaxFileSize = 32000
15          .FileName = ""
16          .ShowSave
17      If Len(.FileName) = 0 Then Exit Sub
```

```
18        If .FileName <> "" And .FileTitle <> "" Then
19          FileName = GetName(.FileName) + "bmp"
20          FileTitle = GetName(.FileTitle) + "bmp"
21
22        If FileExists(FileName) Then
23        Dim Response
24   Response = MsgBox("An image '" + FileTitle + "' already exists in folder " _
25          + CurDir + ". Would you like to replace it ?", vbYesNo, fGUI.Caption)
26        If Response = vbYes Then
27            Obj.FileForm.Tab1.Tab = 1
28            Call obj.FileForm.Tab1_Click(obj.FileForm.Tab1.Tab)
29           SavePicture obj.FileForm.Picture1.Image, GetName(.FileName) + "bmp"
30        End If
31        Else
32            Obj.FileForm.Tab1.Tab = 1
33            Call obj.FileForm.Tab1_Click(obj.FileForm.Tab1.Tab)
34           SavePicture obj.FileForm.Picture1.Image, GetName(.FileName) + "bmp"
35        End If
36      End If
37      End With
38   End If
39   End Sub
        ...
```

After finding the active child window (lines 03-06) using the fact that the attribute "FormID" of an *mdiChild* object is equal to the "FileID" of a *MyFile* object according to lines 05 and 21 in Table 4.4, the old file name and its extension are stored in lines 09-10 followed by the *CommonDialog* object "FileDialog" (lines 11-37) for selecting the wanted image file to store the plot output. If the selected file exists, a response message box is shown. If the user response is positive or the selected file does not exist, the same block of code is executed (lines 27-29 and 32-34). It is **important to set** the "AutoRedraw" attribute of any *PictureBox* object to "True" in the properties window of the VB environment, before a picture can be stored in an image file via the VB procedure "SavePicture". Before the plot needs to be redrawn by calling operation "Tab1_Click", which consists of operations "PlotData" (Table 5.4) and "ShowResult" already mentioned in line 16 of Table 5.6. Notice that only bitmap images can be produced in this way by the VB procedure "SavePicture" so that the extension "bmp" is always added to a displayed and selected file name (19-20). The "GetName" routine in file "DataManager.bas" extracts just the file name without its extension and the dot in the middle.

Table 5.8. Code for "File/Print" in file "GUI.frm" in folder "Program02"

```
     ...
01   Private Sub mnuFilePrint_Click()
```

```
02 | If ChildList.Count > 0 Then
03 |     Screen.MousePointer = vbHourglass
04 |     Dim obj
05 |     For Each obj In ChildList 'Find child in list, then Exit loop
06 |         If Val(Screen.ActiveForm.FormID) = Val(obj.FileID) Then Exit For
07 |     Next obj
08 |
09 |     obj.FileForm.Tab1.Tab = 1
10 |     Set obj.FilePrint = New mdiPrint
11 |
12 |     Dim pWidth As Single, pHeight As Single
13 |     pWidth   =   Printer.ScaleX(Printer.ScaleWidth,   Printer.ScaleMode,   _
14 | vbTwips) * 95 / 100
15 |     pHeight  =   Printer.ScaleY(Printer.ScaleHeight,  Printer.ScaleMode,   _
16 | vbTwips)/ 3
17 |     obj.FilePrint.Picture1.Width = 2 * pWidth
18 |     obj.FilePrint.Picture1.Height = 2 * pHeight
19 |     obj.FilePrint.Width = 2 * pWidth + 100
20 |     obj.FilePrint.Height = 2 * pHeight + 400
21 |
22 |     Call obj.PlotToPrinter(obj.FileForm, obj.FilePrint.Picture1, _
23 |                         2 * pWidth, 2 * pHeight, 1)
24 |     obj.FilePrint.Show
25 |     Screen.MousePointer = vbDefault
26 |
27 |     Dim Response
28 |     Response = MsgBox("Do you want to print this 2D plot of the last active _
29 |                 child window?", vbYesNo, "2D Dynamic Data Plotter")
30 |     If Response = vbYes Then
31 |         Unload obj.FilePrint
32 |     Response = MsgBox("Make sure your printed is switched on and ready!", _
33 |                 vbYes, "2D Dynamic Data Plotter")
34 |         DoEvents
35 |         Call obj.PlotToPrinter(obj.FileForm, Printer, pWidth, pHeight, 3)
36 |         Printer.EndDoc
37 |     Else
38 |         Unload obj.FilePrint
39 |         Printer.KillDoc
40 |         Response = MsgBox("No plot will be printed!", _
41 |                         vbYes, "2D Dynamic Data Plotter")
42 |     End If
43 | End If
44 | End Sub
   |     ...
```

After finding the active child window (lines 05-07) using the fact that the attribute "FormID" of an *mdiChild* object is equal to the "FileID" of a *MyFile* object according to lines 05 and 21 in Table 4.4, a print preview window is created with an *mdiPrint* object named "FilePrint" in line 10 (Fig. 5.2). The attributes of "File-Print" are assigned in lines 12-20 and its *PictureBox* object filled in lines 22-23 by a operation named "PlotToPrinter" with five parameters: frm As Form, obj As Object, Width As Single, Height As Single, DrawWidth As Integer. Then "FilePrint" is displayed (line 24) followed by a response message box where a user decides about sending the 2D plot to the printer or not. If the user response is positive, the plot is sent to the *Printer* object in line 35 and to the physical printer in line 36. Otherwise the print job is deleted (line 39). In any case is object "FilePrint" unloaded (lines 31+38). The important part in this routine is operation "PlotToPrinter" used to create both a print preview (lines 22-23) and the actual print (line 35), but with different arguments to adjust the result to the chosen drawing objects.

5.4 Patterns used in this application

As mentioned in paragraphs 4.1 and 5.1, the **layers architectural pattern** with a user interface, a processor and a data manager is the fundamental three-tier architecture of all applications considered in part II of this book (Fig. 4.1 and 4.2).

Design patterns used in this application are the same ones as in the last application "Program01" plus the Singleton creational pattern:

Creational Patterns:

2. Builder: Separate the construction of a complex object from its representation so that the same construction process can create different representations.

3. Factory Method: Define an interface for creating an object, but let subclasses decide which class to instantiate. Factory Method lets a class defer instantiation to subclasses.

4. Prototype: Specify the kinds of objects to create using a prototypical instance, and create new objects by copying this prototype.

5. Singleton: Ensure a class only has one instance, and provide a global point of access to it.

Behavioral Patterns:

16. Iterator: Provide a way to access the elements of an aggregate object sequentially without exposing its underlying representation.

19. Observer: Define a one-to-many dependency between objects so that when an object changes state, all its dependents are notified and updated automatically.

The **Singleton** pattern is used here when referring to a *Printer* object (Table 5.8) because VB allows only one single *Printer* object. Explanation of the other patterns is identical to paragraph 4.4 and therefore not repeated here.

5.5 Testing the application

The 2D dynamic data plotter was tested with different data text files in various ways in order to fulfill the requirements given in the beginning of paragraph 5.2. These tests can be categorized according to the application's three-tier architecture as follows:
• Testing the user interface,
• Testing the, processor and
• Testing the data manager.

Tests of the user interface and the data manager were not very difficult. It was discovered, for example, that some menu operations like "File/Run" or "File/Print" require an existing *MyFile* object. There is no problem as long as the *ChildList* object for detecting the active child is iterated as written in Table 5.5. But if the *ChildList* loop is exited after finding the active child window object, then the whole routine need to be enclosed into an "If" clause beginning with

02 | If ChildList.Count > 0 Then

as was done in line 02 in Table 5.7 and 5.8. If this code is not included, then a click on such a menu option results in the following error message, if no child window object exists.

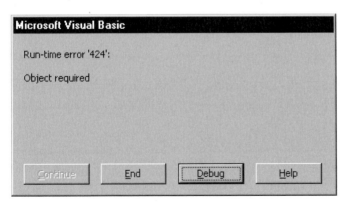

Fig. 5.6. Run-time error message when line 02 in Table 5.7 and 5.8 is exluded and no child window object exists

Most of the testing of this application was done with the processor to compute, display and print a Fourier series according to Equation (5.1) at the end of paragraph 5.2. Although these formulas are only valid for equidistant data points regarding their X coordinate values, they work as well for **non-equidistant data sets** after

1. scaling the data points to equal X distances,
2. build the Fouier analysis, and
3. rescale the X coordinates to their original values.

Exactly this was done in the program considered here as illustrated with the data file called "111 Unequal.txt" in Fig. 5.7.

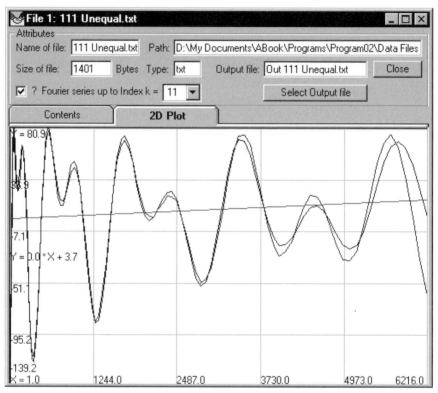

Fig. 5.7. Fourier analysis of a non-equidistant data set "111 Unequal.txt" in folder "Program02/Data Files" (k = 11)

In Fig.5.7, the Fourier index was set to "k = 11" in order to **test which Fourier coefficients are relevant** and which ones can be neglected. They can be viewed together with the differences between the Fourier and the original values after clicking on the "Contents" tab of a child window object, or on "View/Output File" in the program's menu bar. If k is set to the maximum value of 55 for this data set,

Fourier values fit very well to the original values (Fig. 5.8), except at the ends where the series would need to be artificially extended to avoid this problem.

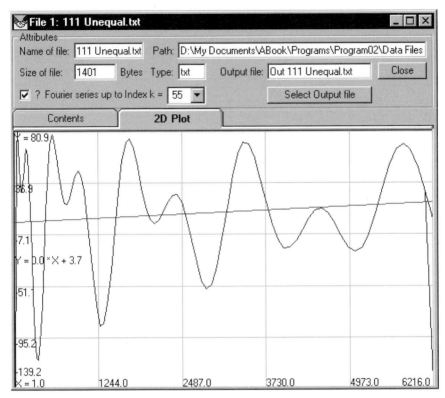

Fig. 5.8. Fourier analysis of a non-equidistant data set "111 Unequal.txt" in folder "Program02/Data Files" (k = 55)

In paragraph 5.3, two options were introduced to export or print a 2D plot:
- Image file outputs by clicking on "File/Save Plot As", or
- Hardcopy printouts by clicking on "File/Print" in the program's menu bar.

The first printing option is easier to program, more effective and gives a better print quality because it can use the printing facilities of your existing word processor, whereas the hardcopy prints of this program are very basic. And text was not included in them because text and graphics are differently scaled on VB *Picture-Box* and *Printer* objects. When this was discovered, the hardcopy printing routine of this program was not further developed because it would require a lot of time and energy to produce sufficient hardcopy quality.

When using the code for the "File/Save Plot As" routine in Table 5.7, it is **important to set** the "AutoRedraw" attribute of any *PictureBox* object to "True" in the properties window of the VB environment, before a picture can be stored in an image file via the VB procedure "SavePicture". An alternative is to insert the following code into Table 5.7 (lines 28 and 34 in Table 5.9), resulting in the table below.

Table 5.9. Alternative code for "File/Save Plot As" in file "GUI.frm" in folder "Program02"

```
        ...
23      Dim Response
24  Response = MsgBox("An image '" + FileTitle + "' already exists in folder " _
25          + CurDir + ". Would you like to replace it ?", vbYesNo, fGUI.Caption)
26      If Response = vbYes Then
27          obj.FileForm.Tab1.Tab = 1
28          obj.FileForm.Picture1.AutoRedraw = True
29          Call obj.FileForm.Tab1_Click(obj.FileForm.Tab1.Tab)
30         SavePicture obj.FileForm.Picture1.Image, GetName(.FileName) + "bmp"
31      End If
32      Else
33          obj.FileForm.Tab1.Tab = 1
34          obj.FileForm.Picture1.AutoRedraw = True
35          Call obj.FileForm.Tab1_Click(obj.FileForm.Tab1.Tab)
36         SavePicture obj.FileForm.Picture1.Image, GetName(.FileName) + "bmp"
37      End If
        ...
```

CHAPTER 6: 3D Data Visualizer in Visual Basic

6.1 Introduction to 3D graphics

The application considered in this chapter follows very much the previous one in terms of the general architecture and how the designed model works in principle. The **main difference** is the incorporation of 3D graphical concepts to be able to visualize the contents of any 3D data file. To make it more interesting, the application is supplemented by an animation utility for time-dependent data sets for watching data changes in movie-style.

Many different ways exist how to visualize 3D data. The most common ones are based on isoclines or **surface polyeders** (triangles, squares, etc.) in combination with some type of **color-coding**. The latter method is also used in this application. Because data points can be discontinuously distributed over a domain, an appropriate interpolation technique is needed to get around this problem. Here a grid of equal-sized squares is firstly computed in the X-Y coordinate plane in order to guarantee an equal data distribution. Then the Z coordinate values of all four-corner points of each square are interpolated, resulting in grid point coordinates Gx,Gy,Gz.

Further, the color value of each square is calculated with the mean Z values of all four-corner points according to the rainbow color code chosen here (Fig. 6.1).

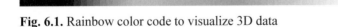

Fig. 6.1. Rainbow color code to visualize 3D data

This color code uses dark blue and dark red to express the minimum and maximum Z coordinate values existing in the given 3D data set. All other colors express values in between.

To watch data in 3D, a **coordinate transformation** is required to translate grid point coordinates Gx,Gy,Gz from the original coordinate system of the given data file to screen coordinates Sx,Sy,Sz. Neglecting rotations around the Z axis for now, the transformation below can be used

$$
\begin{aligned}
Sx &= PosX + ScaleX * (Gx - X0), \\
Sy &= PosY + ScaleY * (Gy - Y0) * \sin \alpha, \\
Sz &= \qquad\quad ScaleY * (Gz - Z0) * \cos \alpha,
\end{aligned}
\tag{6.1}
$$

containing the following parameters:
- Pos X: center of the drawing object (a *PictureBox* object in Fig. 6.3) in x coordinate direction of the screen coordinate system in [pixel],
- Pos Y: center of the drawing object in y coordinate direction in [pixel],
- Scale X: scale factor of the x coordinates,
- Scale Y: scale factor of the y coordinates,
- X0,Y0,Z0 are the mean values of the given data: X0 = (Xmax-Xmin)/2, Y0 = (Ymax-Ymin)/2 and Z0 = (Zmax-Zmin)/2,
- α is the view angle in direction towards the X-Y coordinate plane (Fig. 6.2).

Fig. 6.2. Two rotation angles α and γ determine a view direction

The other angle to determine a view direction is γ, describing rotations around the Z axis. Both angles are used to compute a rotation matrix RM

$$
RM = ROT.x\ (\pi/2 - \alpha) * ROT.z\ (\gamma),
\tag{6.2}
$$

whereby ROT.x and ROT.z are matrices for rotations around coordinate axis given by (Stephens 2000):

$$
ROT.x\ (\alpha) =
\begin{vmatrix}
1 & 0 & 0 \\
0 & \cos \alpha & -\sin \alpha \\
0 & \sin \alpha & \cos \alpha
\end{vmatrix},
$$

$$
ROT.z\ (\gamma) =
\begin{vmatrix}
\cos \gamma & -\sin \gamma & 0 \\
\sin \gamma & \cos \gamma & 0 \\
0 & 0 & 1
\end{vmatrix}.
\tag{6.3}
$$

Applying equation (6.2) to (6.1), one gets the full transformation:

$$Sx = PosX + ScaleX * \{ RM(1,1)*(Gx - X0) + RM(1,2)*(Gy - Y0) \},$$
$$Sy = PosY + ScaleY * \{ RM(2,1)*(Gx - X0) + RM(2,2)*(Gy - Y0) \}, \qquad (6.4)$$
$$Sz = ScaleY * \{ Gz - Z0 \} * \cos \alpha,$$

whereby

$$Sy := Sy + Sz \qquad\qquad\qquad (6.5)$$

is set to a new Sy value because a computer screen is two dimensional and cannot directly display 3D coordinates. Therefore, the z coordinate value in equation (6.4) is added to the y value to produce the final screen coordinate Sy in equation (6.5).

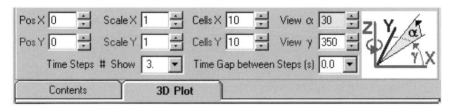

Fig. 6.3. Eight plus two plot parameters to visualize any 3D data

The application considered here consists of eight plus two plot parameters as shown in Fig. 6.3. The first eight parameters are valid for any type of 3D data set, the last two only for time-dependent sets as part of the animation utility. The table below explains each parameter in more detail.

Table 6.1. Eight plus two plot parameters to visualize any 3D data

01	Pos X: center of the drawing object (a *PictureBox* object in Fig. 6.3) in x coordinate direction of the screen coordinate system in [pixel],
02	Pos Y: center of the drawing object in y coordinate direction in [pixel],
03	Scale X: scale factor of the x coordinates,
04	Scale Y: scale factor of the y coordinates,
05	Cells X: number of cells or grid squares in x coordinate direction,
06	Cells Y: number of cells or grid squares in y coordinate direction,
07	View angle α: angle of rotation according to equations (6.2) – (6.5) in [deg],
08	View angle γ: angle of rotation according to equations (6.2) – (6.5) in [deg],
09	Time Steps: number of steps in a given data set,
10	Time Gap: number of seconds until the next step is shown in an animation to display time-dependent data in movie-style.

As can be seen in Fig. 6.3, the first eight plot parameters in Table 6.1 are given through *TextBox* objects each linked to an *UpDown* object on the right-hand side of each *TextBox* object. Both type of objects are provided through the MFC library. *UpDown* objects are easy input controls and parameter selectors because they allow setting the range of applicable parameter values (minima and maxima) as well as the increment. For example, the *UpDown* objects for the view angles α and γ have the following settings:

Min = - 5, Max = 360, Increment = 5,

so that the code below, executed after an *UpDown* object changed its value, guarantees that an angle is always between 0 and 360 deg without stopping at the end values. Thus, there are smooth transitions for values below 0 and larger than 360 deg.

```
01 | If UpDown(6).Value < 0 Then UpDown(6).Value = UpDown(6).Value+ 360
02 | If UpDown(6).Value = 360 Then UpDown(6).Value = 0
03 | If UpDown(7).Value < 0 Then UpDown(7).Value = UpDown(7).Value+ 360
04 | If UpDown(7).Value = 360 Then UpDown(7).Value = 0
```

6.2 Object-oriented analysis and design

The 3D data visualizer considered here should fulfill the following **requirements**. It should allow to:
1. Open 3D input files of variable size,
2. Display a 3D data plot, and
3. Allow to change a 3D data view by variable plot parameters, and
4. Allow an animation for time-dependent data sets.

The class diagram of the 3D data visualizer in Fig. 6.4 is based on the class diagram of the dynamic matrix processor in chapter 4 (Fig. 4.4) plus some other classes added to classes *MyFile* and *mdiChild* (Fig. 6.4). These classes are:
- class *MyPoint* to create 3D data points and their coordinates,
- class *PointList* to manage *MyPoint* objects, and
- class *UpDown* for selecting and controlling variable plot parameters.

All new classes in Fig. 6.4 have a whole-to-part association with classes *MyFile* and *mdiChild*. *MyPoint* is a self-written class with an ID and X,Y,Z coordinates as attributes, whereas *PointList* and *UpDown(i)*, i = 0,1,...,7 for the first eight plot parameters in Table 6.1, have a generalization-specialization association with classes *Collection* and *UpDown* respectively, which belong both to the MFC library.

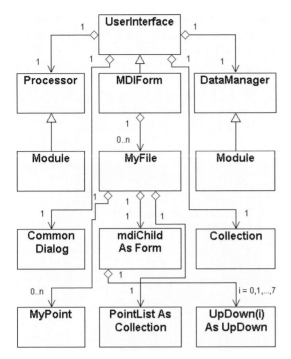

Fig. 6.4. Class diagram for a 3D data visualizer

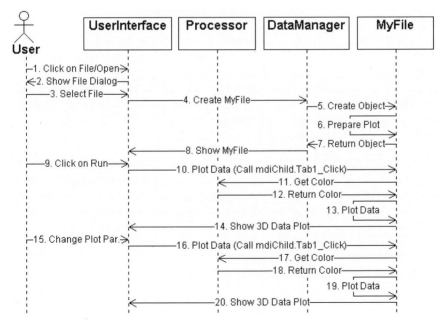

Fig. 6.5. Sequence diagram for a 3D data visualizer

A sequence diagram for the given requirements of this application is given in Fig. 6.5. This diagram is very similar to the sequence diagram of the 2D dynamic data plotter (Fig. 5.3) in chapter 5. After clicking on "Run" in the program's menu bar or on the tab named "3D Plot" of an *SSTab* object (step 9), the 3D data plot is computed and displayed. Apart from the *UpDown* objects for selecting and controlling variable plot parameters (Table 6.1), an *mdiChild* object contains also two *ComboBox* objects for time-dependent data sets to choose the wanted number of time steps ("ALL" or single steps) and the time gap between steps in seconds as shown in Fig. 6.6.

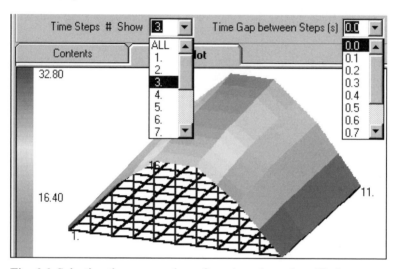

Fig. 6.6. Selecting time step and gap for a time-dependent 3D data set

Both classes *ComboBox* and *SSTab* were not included in the class diagram (Fig. 6.4) to preserve simplicity and readability. The *SSTab* object placed on a child window allows switching back and forth between the digital file output of a *My-File* object (Tab: "Contents") and its analog graphical output (Tab: "3D Plot") by clicking on the corresponding tab (Fig. 6.3 and Fig. 6.6).

The *ComboBox* objects in Fig. 6.6 represent the last two plot parameters in Table 6.1 (09-10) of the animation utility for time-dependent data built into this application to allow users to watch data changes in movie-style. Apart from the chosen time gap the **animation speed** depends very much on the existing computer hardware, for example, the processor speed and type of graphics card. Another important factor influencing the animation speed is the number of cells or grid squares selected by a user. The maximum number for "Cell X" and "Cell Y" is one hundred ("100"), which was found to be a good compromise between display quality and speed.

6.3 Implementation of the designed model

The coded architecture of the 3D data visualizer can be easily visualized with the VB project explorer (McMonnies 2001) as shown in the figure below. This window is automatically displayed after opening the application's project file named "_Program03.vbp" in folder "Program03". All example programs including executable files and source code can be **downloaded** at http://de.geocities.com/ bsttc2/book/SMOP.zip

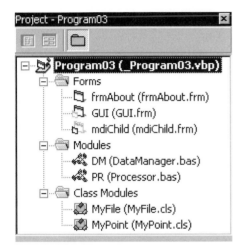

Fig. 6.7. VB project explorer window for the 3D data visualizer

As can be seen in this figure, "Program03" is very similar to "Program02" (Fig. 5.5) with only file "mdiPrint.frm" removed, because the printing facility was deleted in this application because of the reasons given in paragraph 5.5. But the "File/Save Plot As" routine was not changed and can be easily used to produce printouts of any 3D plot.

How "Program03" works and how it was coded, can be more easily described by going through the application's sequence diagram (Fig. 6.5) and explaining step-by-step the corresponding program code. Because the beginning of "Program03" is almost identical to "Program01" and "Program02" until step 8 (Fig. 5.3), a description of this part is not necessary where the contents is identical, which is the case for Tables 4.2, 4.3 and 4.4. The first difference occurs in operation "ReadData" of a *MyFile* object with the following programming code in Table 6.2.

Table 6.2. Code for "ReadData" in file "MyFile.cls" in folder "Program03"

	...
01	Public Sub ReadData()

```
02 | If FileExists(FileName) Then
03 |
04 |    If LCase(FileType) = "txt" Or LCase(FileType) = "dat" Then
05 |       Dim i As Integer, xp As Single, yp As Single, zp As Single, sp As Single
06 |       i = 0
07 |       Set PointList = New Collection   ' Create PointList
08 |       Open FileName For Input As #1 ' Open file for input
09 |          Do While Not EOF(1)          ' Loop until end of file
10 |          i = i + 1
11 |          Input #1, sp, xp, yp, zp
12 |          Dim poi As MyPoint             ' Create new object
13 |          Set poi = New MyPoint          '  of class MyPoint
14 |          poi.ID = Str(sp)
15 |          poi.X = xp
16 |          poi.Y = yp
17 |          poi.Z = zp
18 |          PointList.Add Item:=poi        ' Add poi to PointList
19 |          Set poi = Nothing
20 |          Loop
21 |       Close #1     ' Close file
22 |       Max = i         ' Compute maximum number of data points
23 |
24 |       Call PreparePlot
25 |       Call PrintList(PointList, Max, Steps, "This is the point list (input) :", _
26 |                   FileForm.Text1, OutFName, 0)
27 |    End If
28 |
29 |    If LCase(FileType) = "bmp" Or LCase(FileType) = "gif" _
30 |    Or LCase(FileType) = "jpg" Or LCase(FileType) = "wmf" _
31 |    Or LCase(FileType) = "emf" Then
32 |       FileForm.Picture1.Picture = LoadPicture(FileName)
33 |       FileForm.Tab1.Height = 1170 + FileForm.Picture1.Height
34 |       FileForm.Height = 1600 + FileForm.Tab1.Height
35 |       FileForm.Tab1.Width = FileForm.Width – 135
36 |       FileForm.Picture1.Width = FileForm.Tab1.Width
37 |       FileForm.Tab1.Tab = 1
38 |    End If
39 |
40 | Else
41 |       Max = 0
42 |       FileForm.Tab1.Tab = 0
43 | End If
44 | End Sub
   |    ...
```

After checking the existence of file "FileName" of a *MyFile* object, "If" clauses are used to distinguish different file types in lines 04 and 30-32. Here two types can be read: data text and image files. If a data text file exists, a *Collection* object named "PointList" (a public member class of a *MyFile* object according to Fig. 6.4) is created in line 07 and filled with 3D data points read from file "FileName" (lines 08-21). For doing this, a *MyPoint* object named "poi" is created in lines 12-13, assigned an ID as well as X,Y,Z coodinates (lines 14-17) and then added to object "PointList" in line 18. Then "PointList" is used to print data in the child window area of the user interface and in an output file (lines 25-26). Further, a 3D data plot is prepared by calling an operation "PreparePlot" of a *MyFile* object in line 24.

If the input file is an image file with extensions "bmp", "gif", "jpg", "wmf" or "emf", the image is loaded into a *PictureBox* object named "Picture1" belonging to an *mdiChild* object "FileForm" (line 32). Then objects "Picture1" and the *SSTab* object called "Tab1" are resized (line 33-36) and the second tab of "Tab1" with an index equal to one is made visible (line 37).

If the input file "FileName" of a *MyFile* object does not exist at all, the total number of data points "Max" is set to zero (line 41), and the *SSTab* object "Tab1" to zero (line 42), thus showing the "Contents" tab (Fig. 6.6).

The "PreparePlot" operation in line 24 of Table 6.2 has a code given in the following table. It is used to compute some time-independent plot variables and set up the *ComboBox* objects for the animation utility.

Table 6.3. Code for "PreparePlot" in file "MyFile.cls" in folder "Program03"

```
      ...
01    Private Sub PreparePlot()
02    If FileExists(FileName) Then
03     If LCase(FileType) = "dat" Or LCase(FileType) = "txt" Then
04      If PointList.Count > 0 Then
05
06    Dim poi
07    Dim i As Integer
08    Dim a, x1, y1 As Single
09
10    Set poi = PointList.Item(1)      ' Start initialization
11       Par(1) = poi.X                ' Min for X
12       Par(2) = poi.Y                ' Min for Y
13       Par(3) = poi.Z               ' Min for Z
14       Par(4) = poi.X               ' Max for X
15       Par(5) = poi.Y               ' Max for Y
16       Par(6) = poi.Z               ' Max for Z
17       i = 0
```

```
18        ReDim XX(Max)
19        ReDim YY(Max)
20        ReDim ZZ(Max)
21        x1 = poi.X
22        y1 = poi.Y
23        Steps = 0
24     For Each poi In PointList          ' Compute Minima/Maxima
25       If Par(1) > poi.X Then Par(1) = poi.X
26       If Par(2) > poi.Y Then Par(2) = poi.Y
27       If Par(3) > poi.Z Then Par(3) = poi.Z
28       If Par(4) < poi.X Then Par(4) = poi.X
29       If Par(5) < poi.Y Then Par(5) = poi.Y
30       If Par(6) < poi.Z Then Par(6) = poi.Z
31       i = i + 1
32       XX(i) = poi.X
33       YY(i) = poi.Y
34       ZZ(i) = poi.Z                    ' Check number of Steps
35       If i > 1 And x1 = poi.X And y1 = poi.Y Then Steps = Steps + 1
36     Next poi

38       If Steps > 0 Then               ' Assign step parameter
39          FileForm.Combo1.Clear
40          FileForm.Combo1.AddItem "ALL"
41          For i = 1 To Steps + 1
42             FileForm.Combo1.AddItem Str(i) + "."
43          Next i
44       End If
45          FileForm.Combo1.ListIndex = 0
46          FileForm.Combo2.ListIndex = 0

48       a = 1000
49       p1 = 1      ' Find corner grid points to print their labels: No[p1,p2,p3,p4]
50       For i = 1 To Max    ' Xmin, Ymin   // For the square: 1,6,11,16
51        If dist(XX(i), YY(i), Par(1), Par(2)) < a Then
52          a = dist(XX(i), YY(i), Par(1), Par(2))
53           p1 = i
54        End If
55       Next i
56       a = 1000
57       p2 = 1
58       For i = 1 To Max      ' Xmax, Ymin
59        If dist(XX(i), YY(i), Par(4), Par(2)) < a Then
60          a = dist(XX(i), YY(i), Par(4), Par(2))
61           p2 = i
62        End If
63       Next i
```

```
64    a = 1000
65    p3 = 1
66    For i = 1 To Max     ' Xmax, Ymax
67      If dist(XX(i), YY(i), Par(4), Par(5)) < a Then
68        a = dist(XX(i), YY(i), Par(4), Par(5))
69        p3 = i
70      End If
71    Next i
72    a = 1000
73    p4 = 1
74    For i = 1 To Max     ' Xmin, Ymax
75      If dist(XX(i), YY(i), Par(1), Par(5)) < a Then
76        a = dist(XX(i), YY(i), Par(1), Par(5))
77        p4 = i
78      End If
79    Next i
80
81    End If
82    End If
83    End If
84    End Sub
      ...
```

The code in Table 6.3 for the "PreparePlot" operation of a *MyFile* object is structured as follows. After checking the existence of file "FileName", its extension "FileType" and the number of points in "PointList" (three "If" clauses starting in lines 02-04 and ending in lines 81-83), all needed variables are initialized in lines 10-23 followed by computing the minima and maxima X,Y,Z coordinate values (lines 24-36) stored in "Par(1-6)", which are private member attributes of a *MyFile* object. In line 35, a check is performed to control whether a data set is time-dependent or not. The answer is "Yes", if the X,Y values of the first data point (lines 21-22) appear again inside of the same data set.

In lines 38-46, the two *ComboBox* objects for the animation utility named "Combo1" and "Combo2" are filled and initialized where "FileForm" represents an *mdiChild* object as part of a *MyFile* object. The rest of the "PreparePlot" operation finds the corner grid points to print their labels in a 3D plot and stores the results in parameters "p1" to "p4", which are private member attributes of a *MyFile* object (lines 48-79).

The other steps in the sequence diagram (Fig. 6.5) represent two loops to display (1. loop: steps 9-14) and change a 3D plot (2. loop: steps 15-20), after "Run" in the program's menu bar or the "3D Plot" tab of a *SSTab* object was selected (step 9). By doing this, the code in Table 6.4 and/or 6.5 is executed, whereby the 2. loop is identical to the 1. loop with only different plot parameters (Table 6.1).

Table 6.4. Code for "File/Run" in file "GUI.frm" in folder "Program03"

```
      ...
01  Private Sub mnuRun_Click()
02     Dim obj
03     For Each obj In ChildList   ' Find child in list
04         If val(Screen.ActiveForm.FormID) = val(obj.FileID) Then
05            If obj.FileForm.Tab1.Tab = 0 Then
06                obj.FileForm.Tab1.Tab = 1 ' Plot data
07            Else
08                Call obj.FileForm.Tab1_Click(obj.FileForm.Tab1.Tab)
09            End If
10         End If
11     Next obj
12  End Sub
      ...
```

After finding the active child window using the fact that the attribute "FormID" of an *mdiChild* object is equal to the "FileID" of a *MyFile* object according to lines 05 and 21 in Table 4.4, the "Tab1_Click" operation of an *mdiChild* object with the name "obj.FileForm" is called. The code for this operation is given in Table 6.5.

Table 6.5. "Tab1_Click" in file "mdiChild.frm" in folder "Program03"

```
      ...
01  Public Sub Tab1_Click(PreviousTab As Integer)
02     If Tab1.Tab = 1 Then
03        Dim obj
04        For Each obj In ChildList   ' Find child in list
05           If val(FormID) = val(obj.FileID) Then Exit For
06        Next obj
07           Screen.MousePointer = vbHourglass
08           Call obj.PlotData(obj.FileForm.Picture1)
09           Screen.MousePointer = vbDefault
10     End If
11  End Sub
      ...
```

If "Tab1.Tab" is equal to one ("If" block between lines 02-10), the active *mdiChild* object is detected (lines 04-06), and its "PlotData" operation is called in line 08. The code for this operation is listed in Table 6.6.

Table 6.6. Code for "PlotData" in file "MyFile.cls" in folder "Program03"

```
      ...
01  Public Sub PlotData(obj As Object)
```

```
02    If FileExists(FileName) Then
03        If LCase(FileType) = "dat" Or LCase(FileType) = "txt" Then
04          If PointList.Count > 0 Then
05                          ' Declare plot variables
06          Dim i, j, k, m, n, nx, ny, p, q, t, PosX, PosY As Integer
07          Dim k1, k2, k3 As Integer
08          Dim Color As Long
09          Dim a, rho, scx, scy, Alpha, Gamma As Single
10          Dim Gx(101, 101), Gy(101, 101), Gz(101, 101), Sx(101, 101), _
11              Sy(101, 101), Sz(101, 101), RM(2, 2) As Single
12          Dim x1, y1, x2, y2, x3, y3, x4, y4, x5, y5, x6, y6 As Single
13
14                      ' Assign plot parameters
15          PosX = obj.ScaleX(val(FileForm.Param(0).Text), vbPixels, vbTwips)
16          PosY = obj.ScaleY(val(FileForm.Param(1).Text), vbPixels, vbTwips)
17          Scx = FileForm.UpDown(2).Value / 10
18          Scy = FileForm.UpDown(3).Value / 10
19          nx = val(FileForm.Param(4).Text) + 1
20          ny = val(FileForm.Param(5).Text) + 1
21          Alpha = val(FileForm.Param(6).Text)
22          Gamma = val(FileForm.Param(7).Text)
23
24          rho = 180 / 4 / Atn(1)              ' Compute rotation matrix
25          RM(1, 1) = Cos(Gamma / rho)
26          RM(1, 2) = -Sin(Gamma / rho)
27          RM(2, 1) = Sin(Alpha / rho) * Sin(Gamma / rho)
28          RM(2, 2) = Sin(Alpha / rho) * Cos(Gamma / rho)
29
30                          ' Compute Gx,y,z = grid point coordinates
31          For i = 1 To nx      '  in the original coordinate system
32            For j = 1 To ny    '  of the given data in PointList
33              Gx(i, j) = Par(1) + (i - 1) * (Par(4) - Par(1)) / (nx - 1)
34              Gy(i, j) = Par(2) + (j - 1) * (Par(5) - Par(2)) / (ny - 1)
35            Next j
36          Next i
37
38          For i = 1 To nx     ' Compute Sx,y = screen coordinates for Gx,y
39            For j = 1 To ny
40      Sx(i, j) = PosX + obj.Width / 2 * (1 + scx * (RM(1, 1) * (Gx(i, j) - (Par(4) _
41      + Par(1)) / 2) / (Par(4) - Par(1))+ RM(1, 2) * (Gy(i, j) - (Par(5)+ Par(2))/2) _
42          / (Par(5) - Par(2))))
43
44      Sy(i, j)= PosY + obj.Height / 2 *(0.95+ scy * (RM(2, 1) * (Gx(i, j)-(Par(4) _
45      + Par(1)) / 2) / (Par(4) - Par(1))+ RM(2, 2) * (Gy(i, j) - (Par(5)+ Par(2))/2) _
46          / (Par(5) - Par(2))))
47          Next j
```

```
48        Next i
49
50        obj.Scale (0, obj.Height)-(obj.Width, 0) ' Scale drawing object
51        obj.DrawWidth = 2                        '  and DrawWidth
52
53        If child.Combo1.ListIndex = 0 Then       ' Prepare time loop
54          m = 0
55          n = Steps
56        Else
57          m = child.Combo1.ListIndex – 1
58          n = child.Combo1.ListIndex – 1
59        End If
60          q = Max / (Steps + 1)
61
62        For t = m To n     ' ============= START OF TIME LOOP
63          obj.Cls          ' Clear drawing object
64
65        For i = 1 To nx    ' Compute Gz grid point coordinate (Height)
66        For j = 1 To ny
67          a = 1000
68          k1 = 1
69          k2 = 2
70          k3 = 3
71          For p = t * q + 1 To (t + 1) * q          ' Find 1. Closest data point
72            If dist(XX(p), YY(p), Gx(i, j), Gy(i, j)) < a Then
73              a = dist(XX(p), YY(p), Gx(i, j), Gy(i, j))
74              k1 = p
75            End If
76          Next p
77          a = 1000
78          For p = t * q + 1 To (t + 1) * q          ' Find 2. Closest data point
79            If dist(XX(p), YY(p), Gx(i, j), Gy(i, j)) < a And p <> k1 Then
80              a = dist(XX(p), YY(p), Gx(i, j), Gy(i, j))
81              k2 = p
82            End If
83          Next p
84          a = 1000
85          For p = t * q + 1 To (t + 1) * q          ' Find 3. Closest data point
86        If dist(XX(p), YY(p), Gx(i, j), Gy(i, j)) < a And p <> k1 And p <> k2 Then
87              a = dist(XX(p), YY(p), Gx(i, j), Gy(i, j))
88              k3 = p
89            End If
90          Next p
91                                  ' Solve plane equation in determinant form
92        If det(XX(k2) - XX(k1), YY(k3) – YY(k1), YY(k2) - YY(k1), XX(k3) _
93          - XX(k1)) <> 0 Then
```

94	Gz(i, j) = ZZ(k1) * det(XX(k2) – XX(k1), YY(k3) - YY(k1), YY(k2) _
95	- YY(k1), XX(k3) - XX(k1))
96	Gz(i, j) = Gz(i, j) + (Gy(i, j) - YY(k1)) * det(XX(k2) - XX(k1), ZZ(k3) _
97	- ZZ(k1), ZZ(k2) - ZZ(k1), XX(k3) - XX(k1))
98	Gz(i, j) = Gz(i, j) - (Gx(i, j) - XX(k1)) * det(YY(k2) - YY(k1), ZZ(k3) _
99	- ZZ(k1), ZZ(k2) - ZZ(k1), YY(k3) - YY(k1))
100	Gz(i, j) = Gz(i, j) / det(XX(k2) – XX(k1), YY(k3) - YY(k1), YY(k2) _
101	- YY(k1), XX(k3) - XX(k1))
102	Else ' If on one line, apply line interpolation
103	If (XX(k2) - XX(k1)) <> 0 Then _
104	Gz(i, j) = ZZ(k1) + (ZZ(k2) - ZZ(k1)) / (XX(k2) - XX(k1)) * (Gx(i, j) _
105	- XX(k1))
106	If (YY(k2) - YY(k1)) <> 0 Then _
107	Gz(i, j) = ZZ(k1) + (ZZ(k2) - ZZ(k1)) / (YY(k2) - YY(k1)) * (Gy(i, j) _
108	- YY(k1))
109	End If
110	If Gz(i, j) < Par(3) Then Gz(i, j) = Par(3)
111	If Gz(i, j) > Par(6) Then Gz(i, j) = Par(6)
112	Next j
113	Next i
114	
115	For i = 1 To nx ' Compute Sz = screen coordinates for Gz
116	For j = 1 To ny
117	If (Par(6) – Par(3)) <> 0 Then
118	Sz(i, j) = scy * Cos(Alpha / rho) * (Gz(i, j) – Par(3)) / (Par(6) - Par(3)) _
119	* 2 * obj.Height / 5
120	Else
121	Sz(i, j) = scy * Cos(Alpha / rho) * Par(3) * 2 * obj.Height / 5
122	End If
123	Next j
124	Next i
125	
126	For i = 1 To nx ' Draw horizontal gridlines
127	For j = 1 To ny – 1
128	obj.Line (Sx(i, j), Sy(i, j))-(Sx(i, j + 1), Sy(i, j + 1)), RGB(0, 0, 0)
129	Next j
130	Next i
131	For j = 1 To ny
132	For i = 1 To nx – 1
133	obj.Line (Sx(i, j), Sy(i, j))-(Sx(i + 1, j), Sy(i + 1, j)), RGB(0, 0, 0)
134	Next i
135	Next j
136	
137	i = 1 ' Draw vertical gridlines
138	For j = 1 To ny
139	obj.Line (Sx(i, j), Sy(i, j))-(Sx(i, j), Sy(i, j) + Sz(i, j)), RGB(0, 0, 255)

140	Next j
141	i = nx
142	For j = 1 To ny
143	obj.Line (Sx(i, j), Sy(i, j))-(Sx(i, j), Sy(i, j) + Sz(i, j)), RGB(0, 0, 255)
144	Next j
145	j = 1
146	For i = 1 To nx
147	obj.Line (Sx(i, j), Sy(i, j))-(Sx(i, j), Sy(i, j) + Sz(i, j)), RGB(0, 0, 255)
148	Next i
149	j = ny
150	For i = 1 To nx
151	obj.Line (Sx(i, j), Sy(i, j))-(Sx(i, j), Sy(i, j) + Sz(i, j)), RGB(0, 0, 255)
152	Next i
153	
154	For i = 1 To nx - 1 ' Draw surface cells by
155	For j = 1 To ny - 1 ' filling them with lines
156	x1 = Sx(i, j)
157	y1 = Sy(i, j) + Sz(i, j)
158	x2 = Sx(i + 1, j)
159	y2 = Sy(i + 1, j) + Sz(i + 1, j)
160	x3 = Sx(i + 1, j + 1)
161	y3 = Sy(i + 1, j + 1) + Sz(i + 1, j + 1)
162	x4 = Sx(i, j + 1)
163	y4 = Sy(i, j + 1) + Sz(i, j + 1)
164	a = Sqr((x2 - x1) * (x2 - x1) + (y2 - y1) * (y2 - y1))
165	Color = GetColor((Gz(i, j) + Gz(i + 1, j) + Gz(i + 1, j + 1) + Gz(i, j + _
166	1))/4 - Par(3), Par(3), Par(6))
167	For k = 0 To a Step 30
168	x5 = x1 + k * (x2 - x1) / a
169	y5 = y1 + k * (y2 - y1) / a
170	x6 = x4 + k * (x3 - x4) / a
171	y6 = y4 + k * (y3 - y4) / a
172	obj.Line (x5, y5)-(x6, y6), Color
173	Next k
174	Next j
175	Next i
176	
177	obj.CurrentX = Sx(1, 1) + 20 ' Print corner point labels
178	obj.CurrentY = Sy(1, 1)
179	obj.Print Format(p1, "###0.")
180	obj.CurrentX = Sx(nx, 1) + 20
181	obj.CurrentY = Sy(nx, 1)
182	obj.Print Format(p2, "###0.")
183	obj.CurrentX = Sx(nx, ny) + 20
184	obj.CurrentY = Sy(nx, ny)
185	obj.Print Format(p3, "###0.")

```
186     obj.CurrentX = Sx(1, ny) + 20
187     obj.CurrentY = Sy(1, ny)
188     obj.Print Format(p4, "###0.")
189
190     For i = 1 To obj.Height – 740 Step 26  ' Draw color bar on the left
191         Color = GetColor(i, 0, obj.Height – 740)
192             obj.Line (0, i + 760)-(250, i + 760), Color
193     Next i
194
195     obj.CurrentX = 350                   ' Print Z Min/Max values
196     obj.CurrentY = 980
197     obj.Print Format(Par(3), "###0.00")
198     obj.CurrentX = 350
199     obj.CurrentY = obj.Height / 2 + 500
200     obj.Print Format((Par(3) + Par(6)) / 2, "###0.00")
201     obj.CurrentX = 350
202     obj.CurrentY = obj.Height
203     obj.Print Format(Par(6), "###0.00")
204
205     Dim Gap, Start
206         Gap = val(child.Combo2.Text) + 0.01 ' Set gap in sec
207         Start = Timer                   ' Set start time
208     Do While Timer < Start + Gap
209       DoEvents                          ' Wait time gap
210     Loop
211
212     Next t          ' ============= END OF TIME LOOP
213
214     End If
215     End If
216 End If
217 End Sub
        ...
```

The code in Table 6.5 for the "PlotData" operation of a *MyFile* object is structured as follows. After checking the existence of file "FileName", its extension "FileType" and the number of points in "PointList" (three "If" clauses starting in lines 02-04 and ending in lines 214-216), different blocks of code are sequentially executed to produce a 3D plot. These **blocks** are:

1. Lines 06-13: Declare plot variables,
2. Lines 15-22: Assign plot parameters (Table 6.1),
3. Lines 24-28: Compute rotation matrix according to equations (6.2) – (6.3),
4. Lines 31-36: Compute Gx,y = grid point coordinates,
5. Lines 38-48: Compute Sx,y = screen coordinates of Gx,y according to equations (6.2) – (6.4),

6. Lines 50-51: Scale drawing object and DrawWidth,
7. Lines 53-60: Prepare time loop,
8. Lines 62-212:START-END OF TIME LOOP,
9. Lines 65-113:Compute Gz grid point coordinate (Height),
10. Lines 115-124:Compute Sz = screen coordinate of Gz according to equations (6.2) – (6.5),
11. Lines 126-135:Draw horizontal gridlines,
12. Lines 137-152:Draw vertical gridlines,
13. Lines 154-175:Draw surface cells,
14. Lines 177-188:Print corner point labels,
15. Lines 190-193:Draw color bar on the left,
16. Lines 195-203:Print Z Min/Max values, and
17. Lines 205-210:Set and wait time gap.

The 9. block of code (lines 65-113) to compute the grid point coordinate Gz (Height) uses the plane equation in determinant form given by

$$\begin{vmatrix} x - x1 & y - y1 & z - z1 \\ x2 - x1 & y2 - y1 & z2 - z1 \\ x3 - x1 & y3 - y1 & z3 - z1 \end{vmatrix}, \tag{6.6}$$

where the closest points P1(x1,y1,z1), P2(x2,y2,z2) and P3(x3,y3,z3) of an unknown point P(x,y,z) are used to calculate its z coordinate by solving equation (6.6) for z = Gz, thus applying a linear interpolation function.

Finally, steps 11-12 and 17-18 in the program's sequence diagram (Fig. 6.5) are coded by a "GetColor" routine in lines 165-166. The same routine is also used in line 191 to draw a **color bar** on the left-hand side of a *PictureBox* object inside of an *mdiChild* object (Fig. 6.6). The height of this bar depends on the height of the *PictureBox* object and adjusts automatically when an *mdiChild* object is resized (lines 190-193). The code for the "GetColor" routine is given in the table below.

Table 6.7. "GetColor" in file "Processor.bas" in folder "Program03"

```
      ...
01 Public Function GetColor(ByVal val As Single, ByVal Min As Single, _
02                     ByVal Max As Single) As Long
03    Dim i As Integer, r As Integer, g As Integer, b As Integer, s As Single
04    If (Max - Min) <> 0 Then                    ' Min = blue, Max = red
05        s = val / (Max - Min) * 1080
06        If Abs(s) > 1081 Then s = 0
07        i = s
08       If i >= 0 And i <= 30 Then
09          r = 0
10          g = 0
```

```
11          b = 225 + i
12       End If
13       If i >= 31 And i <= 285 Then
14          r = 0
15          g = i – 30
16          b = 255
17       End If
18       If i >= 286 And i <= 540 Then
19          r = 0
20          g = 255
21          b = 540 – i
22       End If
23       If i >= 541 And i <= 795 Then
24          r = i – 540
25          g = 255
26          b = 0
27       End If
28       If i >= 796 And i <= 1050 Then
29          r = 255
30          g = 1050 – i
31          b = 0
32       End If
33       If i >= 1051 And i <= 1080 Then
34          r = 1305 – i
35          g = 0
36          b = 0
37       End If
38       GetColor = RGB(r, g, b)
39       Else
40       GetColor = RGB(0, 0, 255)
41       End If
42   End Function
        ...
```

This public function of the processor module receives three input parameters in lines 01-02 (val, Min, Max), whereby "val" is normally a number between the "Min" (RGB = 0,0,255 = blue) and "Max" (RGB = 255,0,0 = red) color value. By linear interpolation "val" is assigned an integer value "i" (lines 05-07) to calculate the corresponding color value between "Min" and "Max". The difference "Max" – "Min" is expanded into 1080 rainbow colors and "val" is assigned its corresponding values for Red, Green, Blue (lines 08-38). If the difference "Max" – "Min" is equal to zero checked by an "If" clause starting in line 04 and ending in line 41, the returned color is blue (line 40).

As already mentioned before, the color bar inside of a *PictureBox* object adjusts automatically when an *mdiChild* object is resized. This is also true for the whole

3D plot because every part of it depends on the height and width of the *PictureBox* object as can be seen in lines 38-48 and 115-124 of Table 6.6 where the screen coordinates Sx,y,z are computed. The **resizing** of an *mdiChild* object is managed by the following code.

Table 6.8. "Form_Resize" in file "mdiChild.frm" in folder "Program03"

```
       ...
01 | Private Sub Form_Resize()
02 |       Frame1.Width = Me.Width – 135
03 |       Tab1.Width = Me.Width – 135
04 |       If Me.Height > 1700 Then Tab1.Height = Me.Height – 1700
05 |         Text1.Width = Tab1.Width
06 |         If Tab1.Height > 1080 Then Text1.Height = Tab1.Height – 1080
07 |         Picture1.Width = Tab1.Width
08 |         If Tab1.Height > 320 Then Picture1.Height = Tab1.Height – 320
09 |       If Tab1.Tab = 1 Then
10 |       Dim obj
11 |       For Each obj In ChildList                ' Find child in list
12 |         If val(FormID) = val(obj.FileID) Then    ' Plot Data
13 |           Call obj.FileForm.Tab1_Click(obj.FileForm.Tab1.Tab)
14 |         End If
15 |       Next obj
16 |       End If
17 | End Sub
       ...
```

6.4 Patterns used in this application

As mentioned in paragraphs 4.1 and 5.1, the **layers architectural pattern** with a user interface, a processor and a data manager is the fundamental three-tier architecture of all applications considered in part II of this book (Fig. 4.1 and 4.2). **Design patterns** used in this application are the same ones as in the first application "Program01":

Creational Patterns:

2. Builder: Separate the construction of a complex object from its representation so that the same construction process can create different representations.

3. Factory Method: Define an interface for creating an object, but let subclasses decide which class to instantiate. Factory Method lets a class defer instantiation to subclasses.

4. Prototype: Specify the kinds of objects to create using a prototypical instance, and create new objects by copying this prototype.

<u>Behavioral Patterns:</u>

16. Iterator: Provide a way to access the elements of an aggregate object sequentially without exposing its underlying representation.

19. Observer: Define a one-to-many dependency between objects so that when an object changes state, all its dependents are notified and updated automatically.

The **builder, factory method and prototype patterns** are used in this application when opening files and calling the factory method "CreateMyFile" (Table 4.3 and 4.4). Complex and different *MyFile* objects are created in the same way using the "ReadData" operation, which decides on which subclasses to instantiate (Table 6.2). The **iterator pattern** is being used with the *Collection* object "PointList" (Table 6.3) or "ChildList" (Table 6.4), and the **observer pattern** with the VB "WindowList" attribute of a *Menu* object (Fig 4.8), for example.

6.5 Testing the application

The 3D data visualizer was tested with different data text files in various ways in order to fulfill the requirements given in the beginning of paragraph 6.2. These tests can be categorized according to the application's three-tier architecture as follows:
- Testing the user interface,
- Testing the, processor and
- Testing the data manager.

Tests of the user interface and the data manager were not very difficult because of the gained insights when testing the previous application "Program02" which also deals with data plots, but in 2D. It was discovered, for example, that rescaling an *mdiChild* object causes an error, if its height ("Me.Height" in line 04 in Table 6.8) is smaller than 1700 twips, which is the minimum height of the *SSTab* object "Tab1". To avoid this type of error, "If" clauses had to be inserted in lines 04,06 and 08 in Table 6.8.

Another problem encountered deals with the *UpDown* objects used to select and control values for the Scale X and Scale Y parameters in Table 6.1. Because an *UpDown* object can only be incremented with non-fractional numbers (minimum value = 1, also selected here) while a scale factor requires a number with a fraction, the following code in lines 02-03 has to be added to the "Up-Down_Change" operation of an *UpDown* object:

Table 6.9. Code for "UpDown_Change" in file "mdiChild.frm" in folder "Program03"

```
     ...
01 | Private Sub UpDown_Change(Index As Integer)
02 |      Param(2) = UpDown(2).Value / 10
03 |      Param(3) = UpDown(3).Value / 10
04 | If UpDown(6).Value < 0 Then UpDown(6).Value = UpDown(6).Value + 360
05 |      If UpDown(6).Value = 360 Then UpDown(6).Value = 0
06 | If UpDown(7).Value < 0 Then UpDown(7).Value = UpDown(7).Value + 360
07 |      If UpDown(7).Value = 360 Then UpDown(7).Value = 0
08 |
09 |      If Tab1.Tab = 1 Then Call Tab1_Click(Tab1.Tab)
10 | End Sub
     ...
```

In this way, the *TextBox* objects "Param(2)" and "Param(3)" of Scale X and Scale Y always show correct values like "1.1" for 11/10 or "0.9" for 9/10. It is important to have this numbers with a dot! But some Windows systems can display these numbers with a comma, such as "1,1" or "0,9" which leads to errors, if the parameter assignment for Scale X and Scale Y (lines 17-18 in Table 6.6) is done with *TextBox* objects like

```
17 | Scx = val(FileForm.Param(2).Text)
18 | Scy = val(FileForm.Param(3).Text)
```

This is the reason why the following lines of code are used in Table 6.6.

```
17 | Scx = FileForm.UpDown(2).Value / 10
18 | Scy = FileForm.UpDown(3).Value / 10
```

Another way to change plot parameters in this program is to insert a parameter value directly into the corresponding *TextBox* object, for example, "Param(2)" or "Param(3)" for Scale X or Scale Y, respectively. When the *TextBox* object is left by the user and thus looses the focus of the application, the inserted value is checked and updated, if it is within the allowed range. This functionality is illustrated with the code in the following table.

Table 6.10. . Code for "Param_LostFocus" in file "mdiChild.frm" in folder "Program03"

```
     ...
01 | Private Sub Param_LostFocus(Index As Integer)
```

```
02      Dim Text
03          Text = Param(Index).Text
04          If Index = 2 Or Index = 3 Then
05              If val(Text) < 1 / 10 Then
06                  MsgBox "A scale value has to be equal to 0.1 or larger!", _
07                          VbOKOnly, "Correct your input"
08                  If Index = 2 Then
09                      Param(Index).Text = Str(1 / 10)
10                      UpDown(2).Value = 1
11                  End If
12                  If Index = 3 Then
13                      Param(Index).Text = Str(1 / 10)
14                      UpDown(3).Value = 1
15                  End If
16              Else
17                  If Index = 2 Then UpDown(2).Value = val(Text) * 10
18                  If Index = 3 Then UpDown(3).Value = val(Text) * 10
19              End If
20          End If
21
22          If Index = 4 Or Index = 5 Then
23              If val(Text) < 1 Or val(Text) > 100 Then
24                  MsgBox "A cell number has to be in the range of 1 to 100!", _
25                          vbOKOnly, "Correct your input"
26                  If Index = 4 Then
27                      Param(Index).Text = "10"
28                      UpDown(4).Value = 10
29                  End If
30                  If Index = 5 Then
31                      Param(Index).Text = "10"
32                      UpDown(5).Value = 10
33                  End If
34              Else
35                  If Index = 4 Then UpDown(4).Value = val(Text)
36                  If Index = 5 Then UpDown(5).Value = val(Text)
37              End If
38          End If
39
40  End Sub
        ...
```

This operation "Param_LostFocus" consists of two blocks of code:
- Lines 04-20 for "Param(2)" and "Param(3)": Scale X,Y: scale factors of the x,y coordinates, and
- Lines 22-38 for "Param(4)" and "Param(5)": Cells X,Y: number of cells or grid squares in x,y coordinate direction.

As mentioned before regarding Table 6.9, *TextBox* objects "Param(2)" and "Param(3)" are a little different from the other plot parameters due to the fractional part of their values. Apart from this, both blocks are almost identical. The main differences are the "If" clauses in lines 05 and 23 for checking the allowed range and the text of the corresponding message boxes (lines 06-07 and 24-25), if the check result is true, meaning that an inserted value lies not within the acceptable range. Otherwise the linked *UpDown* object is updated (lines 17-18 and 35-36). Generally speaking, the code in Table 6.10 demonstrates how user input can be easily controlled and corrected with the help of *TextBox* and linked *UpDown* objects.

Most of the test time of this application was spent on the "PlotData" operation of an *mdiChild* object. Various tests on different machines have shown that this operation is very slow on less powerful PCs with a CPU speed of 133 MHz or slower. The main reason for this is the self-written code to fill a non-rectangular polygon of a surface cell with a given color (lines 154-175 in Table 6.6) because the VB version 5 does not provide such a routine. The identical "PlotData" operation written in Delphi runs about twice as fast on the same computer because Delphi provides a "Polygon" routine that can be filled with a wanted color and style, as shown in the following **Delphi** example.

```
01 | var
02 | poly: array[1..4] of TPoint;
03 | begin
04 | poly[1].X := 200; poly[1].Y := 200;
05 | poly[2].X := 300; poly[2].Y := 300;
06 | poly[3].X := 200; poly[3].Y := 400;
07 | poly[4].X := 100; poly[4].Y := 300;
08 | Image1.Canvas.Pen.Color := clBlue;
09 | Image1.Canvas.Pen.Width := 3;
10 | Image1.Canvas.Brush.Color := clRed;
11 | Image1.Canvas.Brush.Style := bsSolid;
12 | Image1.Canvas.Polygon(poly);
13 | end;
```

This example draws a **tilted square** colored in red with blue edges on the canvas of a *TImage* object named "Image1" (not possible in VB version 5).

The operation "PlotData" in Table 6.6 contains another weakness that explains a **slower speed** because the three closest grid points of each cell are re-calculated for each time step (lines 67-90), though they are constant because the grid point coordinates Gx,Gy are time-independent. Therefore, this part can be moved from the time loop to another convenient place, for example, a *MyFile* operation called "PrepareGrid" listed in Table 6.11. Because the closest grid points depend on the grid point coordinates, Gx,Gy must also be jointly moved and computed in the beginning of the "PrepareGrid" operation. Consequently, Gx,Gy can no longer be declared in "PlotData", but have to be defined in the *MyFile* declarations so that they can be used by both, "PrepareGrid" and "PlotData". Therefore, moving the following blocks of code in Table 6.6 can create a faster version of the considered application:

- Lines 31-36:Compute Gx,y = grid point coordinates, and
- Lines 67-90:Calculate the three closest grid points for computing Gz grid point coordinates (Heights).

These lines are moved from operation "PlotData" to "PrepareGrid", which is called at the end of operation "PreparePlot", or after selecting a new number of Cells X,Y. This functionality is inserted after line 07 of the "UpDown_Change" operation in Table 6.9 before "PlotData" is called (see "UpDown_Change" in file "mdiChild.frm" in folder "Program03/Faster Version"). As is known from numerical methods, a re-meshing of a grid is always very time-consuming. But the faster version described here increases the **program performance** by about 20 % when other plot parameters than Cells X or Cells Y are changed.

Table 6.11. Code for "PrepareGrid" in file "MyFile.cls" in folder "Program03/Faster Version"

```
     ...
01 | Public Sub PrepareGrid()
02 | If FileExists(FileName) Then
03 |     If LCase(FileType) = "dat" Or LCase(FileType) = "txt" Then
04 |         If PointList.Count > 0 Then
05 |
06 |         Dim i, j, nx, ny, p, q, k1, k2, k3 As Integer
07 |         Dim a As Single
08 |
09 |         nx = val(FileForm.Param(4).Text) + 1
10 |         ny = val(FileForm.Param(5).Text) + 1
11 |         ReDim CloGridPoint(nx, ny, 3)
12 |         ReDim Gx(nx, ny)
13 |         ReDim Gy(nx, ny)
14 |
15 |                     ' Compute Gx,y,z = grid point coordinates
16 |         For i = 1 To nx      '  in the original coordinate system
17 |          For j = 1 To ny    '   of the given data in PointList
18 |            Gx(i, j) = Par(1) + (i - 1) * (Par(4) - Par(1)) / (nx - 1)
19 |            Gy(i, j) = Par(2) + (j - 1) * (Par(5) - Par(2)) / (ny - 1)
20 |          Next j
21 |         Next i
22 |
23 |         q = Max / (Steps + 1)
24 |         For i = 1 To nx       ' Compute the three closest data points
```

```
25        For j = 1 To ny
26          a = 1000
27          k1 = 1
28          k2 = 2
29          k3 = 3
30          For p = 1 To q ' Find 1. closest data point
31            If dist(XX(p), YY(p), Gx(i, j), Gy(i, j)) < a Then
32              a = dist(XX(p), YY(p), Gx(i, j), Gy(i, j))
33                k1 = p
34            End If
35          Next p
36          a = 1000
37          For p = 1 To q ' Find 2. closest data point
38            If dist(XX(p), YY(p), Gx(i, j), Gy(i, j)) < a And p <> k1 Then
39              a = dist(XX(p), YY(p), Gx(i, j), Gy(i, j))
40                k2 = p
41            End If
42          Next p
43          a = 1000
44          For p = 1 To q ' Find 3. closest data point
45      If dist(XX(p), YY(p), Gx(i, j), Gy(i, j)) < a And p <> k1 And p <> k2 Then
46              a = dist(XX(p), YY(p), Gx(i, j), Gy(i, j))
47                k3 = p
48            End If
49          Next p
50          CloGridPoint(i, j, 1) = k1
51          CloGridPoint(i, j, 2) = k2
52          CloGridPoint(i, j, 3) = k3
53        Next j
54      Next i
55
56      End If
57      End If
58      End If
59      End Sub
        ...
```

The code of this *MyFile* operation "PrepareGrid" is just taken from operation "PlotData" (lines 31-36 and 67-90 in Table 6.6). The results for the three closest grid points are stored in a new plot variable "CloGridPoint" declared as a private attribute of a *MyFile* object and then used again in "PlotData" after line 67 of Table 6.6.

6.6 Data animation examples

The examples in this paragraph demonstrate the usefulness of data animations and simulations for certain applications. The first example shows the cooling process of a quadratic piece of material. It cools from an initial temperature of Z = T = 30°C down to Z = T = 0°C (Friedrich 1999). The results are displayed in Fig. 6.8.

Fig. 6.8. Diffusion process of a quadratic piece of material, made with file "14 Steps Square.txt" in folder "Program03/Data Files" (Cells X,Y = 30)

As visible in this figure, the highest temperature is always in the geometrical center of the material whereas on the edges, the temperature is the smallest.

The other examples considered here deal with the detection of strongly eroded archaeological signs (Friedrich 1999). The goal of these animations is to simulate the shape degradation of hieroglyphs and try to match their remains to existing signs that are known form other sources. Hieroglyphs are signs made of pictures similar to letters in the Chinese alphabet.

Thinking of hieroglyphic degradation and replacing the temperature in Fig. 6.8 by a hieroglyph's surface height, Fig. 6.8 can be interpreted as follows: a square-shaped hieroglyph looses first its round corners and deforms then into a circular shape because material is diffusing radially based on the 2D diffusion equation

$$\partial^2 u / \partial x^2 + \partial^2 u / \partial y^2 = (1/K)\,(\partial u / \partial t), \tag{6.7}$$

where u = u(,x,y) is the unknown function (here u = surface height = Z), and K is the material constant. A larger K or softer material causes a faster degradation, which finishes after all material disappears, leaving a completely flat surface with zero heights everywhere inside of the original boundary. Thus, K has only an effect on the degradation speed of a hieroglyph but not on the geometrical changes of its shape, which only depends on the boundary geometry at the beginning of the degradation process.

Another hieroglyph examined here is a bird with spread wings. Its degradation process is shown in the following series of pictures (Fig. 6.9).

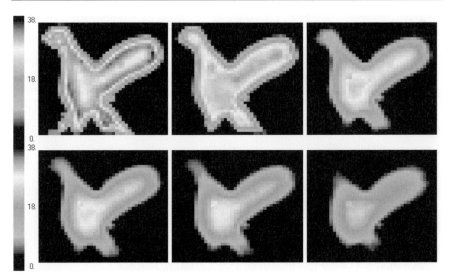

Fig. 6.9. Degradation of a bird with spread wings, made with file "10 Steps Bird.txt" in folder "Program03/Data Files" (Cells X,Y = 40)

The evaluation of Fig. 6.9 gives the following results:

- Sharp corners of hieroglyphs are rounded off.
- Lengthy parts of hieroglyphs become thinner.
- Thinner parts of hieroglyphs deteriorate faster than thicker parts, for example, the head, legs or tail.
- The average surface height of hieroglyphs decreases in time.

CHAPTER 7: Interactive Window Shell for Exe-Files in Visual Basic

7.1 Introduction to updating legacy software

This chapter describes a method to integrate executable files of legacy programs into a visual interactive window shell. Such a shell has the advantage that the migration of legacy programs from e.g. C, Fortran or Pascal to another programming language can be avoided, so that investments in these programs are retained. It is a hybrid approach that allows combining the qualities of legacy code with the visual interactivity of window programs.

Legacy software are computer programs that are difficult to maintain and incompatible with current state-of-the-art operating systems like MS-DOS executables running under Windows XP (Coyle 2000). What can be done with these programs? There are two possible solutions, either program migration or integration (Coyle 2000). Because code migration to another programming language is very time-consuming, error-prone and wasting a lot of the previous investments, it can be more advantageous to integrate legacy programs into another suitable computer environment. How this can be done is illustrated with the shell program in this chapter. Such an approach to integrate legacy executable files into a visual interactive shell offers a lot of **advantages**, for example:

1. Legacy programs need not to be re-written in another language.
2. Data input and output can be done easier and more flexible with window objects and controls.
3. Run-time information from legacy programs can be continuously obtained through ongoing data file exchange.
4. Data output can be immediately visualized and interpreted with the help of graphical components.
5. Results of different legacy programs and data sets created for the same purpose can be easily compared and evaluated.
6. Such a shell program is a convenient tool to collect, select and manage all types of executable files and data sets.

A visual interactive shell is also very beneficial for computational experimentation to evaluate different models. The proposed shell program can be used for this purpose because it easily allows verifying different data sets and models provided by a legacy program. Such verification can be based on:

- comparing results of different data sets using the same model,
- comparing results of different models but using the same data set, and
- comparing results of different data sets and different models.

These different evaluation options are very useful for checking and **calibrating models** when, for example, no correct solution exists. In such a case, different combinations of data sets and models can be tested and, if one of them contains a large difference in data or model, the comparison of different combinations of data sets and models should be able to show it. For evaluating them, data plots of encountered differences are very helpful and are included as well.

Further, the proposed shell program can be used as a **tool for error estimation** and sensitivity studies by varying one or more input parameters and comparing the corresponding results. This can be realized by a **multi-file system** that allows opening and managing different data text files for in- and output at the same time. Its main items and their description are listed in the following Table 7.1.

Table 7.1. Multi-file system for in- and output files

Item	Description
File name definitions (hard coded in "My-Name.EXE")	"MyName.EXE" reads "MyName.TXT" which contains its in- and output file named "MyName.INP" and "MyName.OUT".
Clicks on "File/Open" selecting the same file "MyName.EXE"	1st file opened is named "1MyName.TXT", 2nd file opened is named "2MyName.TXT", 3rd file opened is named "3MyName.TXT", ... n^{th} file opened is named "nMyName.TXT", containing "nMyName.INP" and "nMyName.OUT", which can be also changed and edited at run-time.
Click on "File/Run"	1. Copy "MyName.TXT" to "OldMyName.TXT", 2. Copy "nMyName.TXT" to "MyName.TXT", 3. Call "MyName.EXE" which reads "nMyName.INP" and produces "nMyName.OUT", 4. Copy "OldMyName.TXT" to "MyName.TXT", 5. Show "nMyName.OUT" As Text and Plot, 6. If n >= 2 Then Show "n1MyName.OUT" - "n2MyName.OUT" As Text and Plot.

This multi-file system can be very easily realized by using the parent / multiple child object system of the previous applications in chapters 04-06 with a **file as the main building block** (*MyFile* as the base class).

The window shell chosen here is based on the Microsoft Windows operating system, and the programming language to implement the designed shell is Visual Basic (VB). The final shell program was tested with different types of executable files for:

- Deformation analysis of geodetic networks,
- Earth satellite orbit determination, and
- Tidal acceleration computation.

7.2 Object-oriented analysis and design

The interactive window shell for executable files considered here should fulfill the following **requirements**. It should allow to:

1. Call, run and work with different executable files at the same time,
2. Edit and control data in- and output visually and interactively at run-time, and
3. Visualize results and differences between comparable programs and/or data files.

The flow chart of a legacy program with an executable file (exe-file) in between data input and output (I/O) files is shown in the following figure.

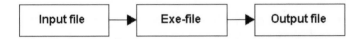

Fig. 7.1. Flow chart of a legacy program with I/O files

An exe-file reads an input file, processes its data and prints the results in an output file. On top of this, a window shell is placed which is connected to exe-and I/O files as given in Fig. 7.2.

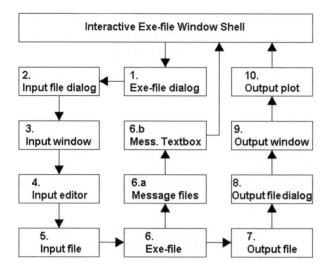

Fig. 7.2. Flow chart of the exe-file shell program

This flow chart in Fig. 7.2 can be easily described through the following use-case diagram.

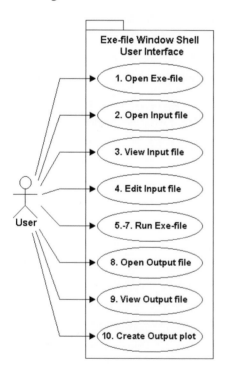

Fig. 7.3. Use-case diagram for the exe-file shell program with 1-10 steps

This use-case can be applied to any executable file that uses I/O files as shown in Fig. 7.1. These I/O files have to be opened through file dialogs by manual selection (steps 2+8 in Fig. 7.3). The exe-file cannot use other files, thus works in single mode (Mode 1). But if the multi-file system in Table 7.1 was coded in an exe-file, the shell program can work in multiple mode (Mode 2) where steps 2+8 are done automatically and different I/O files can be opened and compared, if the same exe-file is opened several times.

Table 7.2. Working modes of the exe-file shell program

Mode	Description
1. Single	Multi-file system (Table 7.1) **not** coded; Steps 2+8: manual
2. Multiple	Multi-file system (Table 7.1) coded; Steps 2+8: automatic

Further, the exe-file can communicate with the shell program during run-time through message files created by the exe-file, if coded before (Fig. 7.2), which can be read and displayed by the shell program in message text boxes so that one can see what an exe-file is doing during run-time. This functionality is optional and described here as multiple mode with run-time file messages (Mode 3).

The class diagram of the exe-file shell program in Fig. 7.4 is based on the class diagram of the dynamic matrix processor in chapter 4 (Fig. 4.4) plus some other classes added to classes *MyFile* and *mdiChild* (Fig. 7.4), which are:

- class *MyPoint* to create data points and their coordinates,
- class *PointList* to manage *MyPoint* objects, and
- class *CheckBox* for selecting and controlling plot parameters.

In Fig. 7.4, all new classes have a whole-to-part association with classes *MyFile* and *mdiChild*. *MyPoint* is a self-written class with an ID and Xi,Yi coordinates as attributes, whereas *PointList* and *CheckBox(i)*, i = 0,1,...,10, have a generalization-specialization association with classes *Collection* and *CheckBox* respectively, which belong both to the MFC library. To avoid information overloading in class *mdiChild*, the *CheckBox(i)* class has not been directly integrated into *mdiChild*, but through another class *frmSelect* which is also a specialized *Form* class like *mdiChild*.

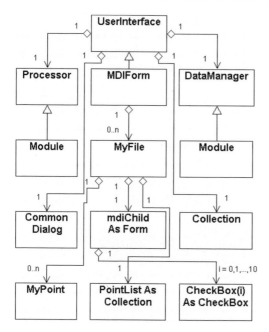

Fig. 7.4. Class diagram for the exe-file shell program

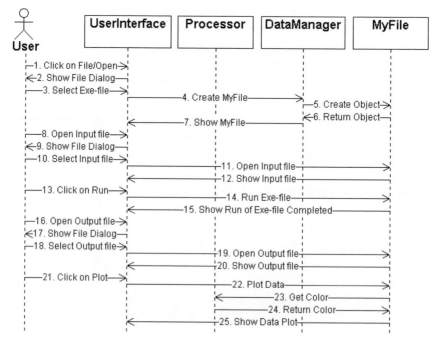

Fig. 7.5. Sequence diagram for the exe-file shell program

A sequence diagram for the given requirements of this application and the use-case diagram in Fig. 7.3 is given in Fig. 7.5. This diagram is very similar to the sequence diagram of the 2D dynamic data plotter (Fig. 5.3) in chapter 5. One major difference is that the *SSTab* object of an *mdiChild* window contains here three tabs named "Input", "Output" and "Plot" (Fig. 7.6). Each of these tabs is linked to the following tasks. A click on:

- Tab "Input" shows the contents of a data input file read by an exe-file.
- Tab "Output" shows the contents of a data output file computed by the exe-file,
- Tab "Plot" displays the contents of a data output file as 2D plot, if the data is given in a table format like in the output window of Fig. 7.6.

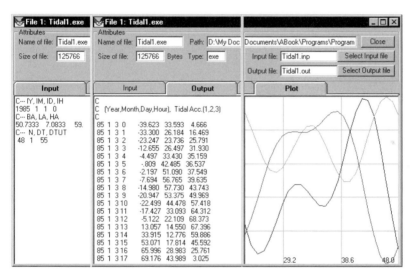

Fig. 7.6. An *mdiChild* object with an *SSTab* object of the exe-file shell program

In Fig. 7.6, the output file consists of 7 data columns but the plot displays only the last three columns. A user selects them before the plot is shown through *CheckBox(i)* objects as part of a *frmSelect* object as in Fig. 7.7.

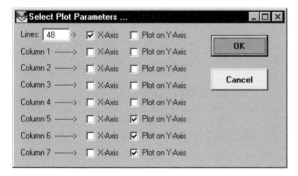

Fig. 7.7. *CheckBox(i)*, i = 0,1,...,10, objects as part of a *frmSelect* object

7.3 Implementation of the designed model

The coded architecture of the exe-file shell program can be easily visualized with the VB project explorer (McMonnies 2001) as shown in the figure below. This window is automatically displayed after opening the application's project file named "_Program04.vbp" in folder "Program04". All example programs including executable files and source code can be **downloaded** at http://de.geocities.com/ bsttc2/book/SMOP.zip

Fig. 7.8. VB project explorer window for the exe-file shell program

As can be seen in this figure, "Program04" is very similar to "Program03" (Fig. 6.7) with only file "mdiSelect.frm" added to display Fig. 7.7.

How "Program04" works and how it was coded, can be more easily described by going through the application's sequence diagram (Fig. 7.5) and explaining step-by-step the corresponding program code. Because the beginning of "Program04" is almost identical to "Program01" and "Program02" until step 8 (Fig. 5.3), a description of this part is not necessary where the contents is identical, which is the case for Tables 4.2, 4.3 and 4.4. Further, steps 8-12 and 16-20 in Fig. 7.5 to open I/O files will not be further explained here because they work in the same way as steps 1-3 described in Table 4.3. The first difference occurs in operation "ReadData" of a *MyFile* object with the following programming code listed in Table 7.3.

Table 7.3. Code for "ReadData" in file "MyFile.cls" in folder "Program04"

	...
01	Public Sub ReadData()

```
02 | If FileExists(FileName) Then
03 |     Dim s As String, nl As String, inp As String, out As String
04 |     nl = Chr(13) & Chr(10)
05 |
06 |     If LCase(FileType) = "exe" Then
07 |       If FileExists(GetName(FileName) + "txt") Then ' Mode 2+3 Exe-files
08 |       Open GetName(FileName) + "txt" For Input As #1 ' Read I/O file names
09 |         Line Input #1, inp
10 |         Line Input #1, out
11 |       Close #1    ' Close file
12 |                       ' Assign in- and output file attributes
13 |       InpFTitle = Right(FileID, Len(FileID) - 1) + inp ' If too long, shorten it
14 |       If Len(InpFTitle) > 12 Then InpFTitle = Left(InpFTitle, 8) + ".inp"
15 |       InpFPath = FilePath
16 |       InpFName = InpFPath + "\" + InpFTitle
17 |       If FileExists(InpFName) = False Then FileCopy inp, InpFTitle
18 |       FileForm.Text(4) = InpFTitle
19 |
20 |       OutFTitle = Right(FileID, Len(FileID) - 1) + out ' If too long, shorten it
21 |       If Len(OutFTitle) > 12 Then OutFTitle = Left(OutFTitle, 8) + ".out"
22 |       OutFPath = FilePath
23 |       OutFName = OutFPath + "\" + OutFTitle
24 |       FileForm.Text(5) = OutFTitle
25 |
26 |       If FileExists(InpFName) Then
27 |       Open InpFName For Input As #1   ' Open file for input
28 |         Do While Not EOF(1)           ' Loop until end of file
29 |         Line Input #1, s             ' Display it in input window
30 |         FileForm.Text1.Text = FileForm.Text1.Text + s + nl
31 |         Loop
32 |       Close #1     ' Close file
33 |       End If
34 |       End If
35 |     End If
36 |
37 |     If LCase(FileType) = "txt" Or LCase(FileType) = "dat" Then
38 |       Open FileName For Input As #1 ' Open file for input
39 |         Do While Not EOF(1)        ' Loop until end of file
40 |         Line Input #1, s
41 |         FileForm.Text1.Text = FileForm.Text1.Text + s + nl
42 |         Loop
43 |       Close #1     ' Close file
44 |     End If
45 |
46 |     If Lcase(FileType) = "bmp" Or Lcase(FileType) = "gif" _
47 |       Or LCase(FileType) = "jpg" Or Lcase(FileType) = "wmf" _
```

```
48 |    Or LCase(FileType) = "emf" Then
49 |       FileForm.Picture1.Picture = LoadPicture(FileName)
50 |       FileForm.Tab1.Height = 440 + FileForm.Picture1.Height
51 |       FileForm.Height = 1600 + FileForm.Tab1.Height
52 |       FileForm.Tab1.Width = FileForm.Width – 135
53 |       FileForm.Picture1.Width = FileForm.Tab1.Width
54 |       FileForm.Tab1.Tab = 2
55 |    End If
56 |
57 | Else
58 |       Max = 0
59 |       FileForm.Tab1.Tab = 0
60 | End If
61 | End Sub
   |       ...
```

After checking the existence of file "FileName" of a *MyFile* object, "If" clauses are used to distinguish different file types in lines 06, 37 and 46-48. Here three types can be read: exe-, data text and image files. If an exe-file of Mode 2 or 3 exists (line 07) according to Table 7.1 and 7.2, its I/O file names are read (lines 08-11) and the corresponding file attributes are assigned in lines 13-18 and 20-24, respectively. Then the input file is opened and its data displayed in the input window, a *TextBox* object named "Text1" belonging to an *mdiChild* object "File-Form".

If file "FileName" is a data text file with extensions "txt" or "dat", it is also opened and its contents displayed in the input window of an *mdiChild* object (lines 37-44). If the selected file is an image file with extensions "bmp", "gif", "jpg", "wmf" or "emf", the image is loaded into a *PictureBox* object named "Picture1" as part of an *mdiChild* object "FileForm" (line 49). Then objects "Picture1" and the *SSTab* object called "Tab1" are resized (line 50-53) and the third tab of "Tab1" with an index equal to two is made visible (line 54). If file "FileName" does not exist, the maximum number of output data file lines called "Max" is set to zero (line 58) and the input window is made visible (line 59).

At this point, step 12 of the sequence diagram (Fig. 7.5) is completed and in step 13, the user clicks on "File/Run" in the program's menu bar to run the exe-file. This is done by the following code given in Table 7.4

Table 7.4. Code for "File/Run" in file "GUI.frm" in folder "Program04"

```
   |    ...
01 | Private Sub mnuRun_Click()
02 |       Dim obj
03 |       For Each obj In ChildList ' Find child in list
04 |          If val(Screen.ActiveForm.FormID) = val(obj.FileID) Then
```

```
05              Screen.MousePointer = vbHourglass
06              Call obj.RunExeFile
07              Screen.MousePointer = vbDefault
08          End If
09      Next obj
10  End Sub
        ...
```

After finding the active child window using the fact that the attribute "FormID" of an *mdiChild* object is equal to the "FileID" of a *MyFile* object according to lines 05 and 21 in Table 4.4, the "RunExeFile" operation of a *MyFile* object is called. The code for this operation is given in Table 7.5.

Table 7.5. "RunExeFile" in file "MyFile.cls" in folder "Program04"

```
    ...
01  Public Sub RunExeFile()
02  If FileExists(FileName) And LCase(FileType) = "exe" Then ' Any Exe-File
03      If FileExists(InpFName) And FileForm.Text1 <> "" Then      ' Mode 1
04        If FileExists(GetName(FileName) + "txt") Then                ' Mode 2+3
05          Call FileForm.Text_LostFocus(4) ' Update InpFTitle and contents
06          Call FileForm.Text_LostFocus(5) ' Update OutFTitle and contents
07          ChDir (FilePath)      ' Go to FilePath directory and backup FileName.txt
08          FileCopy GetName(FileName) + "txt", GetName(FileName) + "tx1"
09          Open GetName(FileName) + "txt" For Output As #1
10            Print #1, InpFTitle
11            Print #1, OutFTitle
12          Close #1
13          Open InpFName For Output As #1  ' Update input files
14            Print #1, FileForm.Text1
15          Close #1
16          Open FilePath + "\" + InpFTitle For Output As #1 ' Temporary input file
17            Print #1, FileForm.Text1
18          Close #1
19
20          If FileExists(FilePath + "\Ok.end") Then Kill FilePath + "\Ok.end"
21          Dim Start
22          Start = Timer
23          Call FileForm.StartBar          ' Start progress bar
24          Dim o, i As Integer, f As String, s As String
25          i = 1
26          o = Shell(FileName, vbNormalFocus) ' Run shell program
27          Do While FileExists(FilePath + "\ok.end") = False And Timer < Start + 20
28            DoEvents                 ' Run progress bar
29            If Timer - Start > 20 Then Start = Start + 20
30            fGUI.ProgressBar1.Value = val(ValToStr(Str(Timer - Start), -1))
```

```
31        f = FilePath + "\" + Right(Str(i), Len(Str(i)) - 1) + ".msg"
32        If FileExists(f) Then
33           Open f For Input As #1
34           Line Input #1, s
35           FileForm.ProgressText.Text = s
36           Close #1
37           i = i + 1
38           Kill f
39        End If
40     Loop
41
42     Call FileForm.StopBar      ' Re-copy FileName.txt
43     If FileExists(GetName(FileName) + "tx1") Then
44        FileCopy GetName(FileName) + "tx1", GetName(FileName) + "txt"
45        Kill GetName(FileName) + "tx1"
46     End If                     ' Copy OutFName file
47     If FileExists(FilePath + "\" + OutFTitle) And _
48        FilePath + "\" + OutFTitle <> OutFName Then _
49        FileCopy FilePath + "\" + OutFTitle, OutFName
50     If FileExists(FilePath + "\" + InpFTitle) And _  'Delete temporary I/O files
51        FilePath + "\" + InpFTitle <> InpFName Then _
52        Kill FilePath + "\" + InpFTitle
53     If FileExists(FilePath + "\" + OutFTitle) And _
54        FilePath + "\" + OutFTitle <> OutFName Then _
55        Kill FilePath + "\" + OutFTitle
56     FileForm.Tab1.Tab = 1
57   Else
58                              ' Mode 1 Exe-file
59     Open InpFName For Output As #1  ' Update input file
60        Print #1, FileForm.Text1
61     Close #1
62     o = Shell(FileName, vbNormalFocus) ' Run shell program
63     If FileForm.Text(5).Text = "" Then Call FileForm.Command3_Click
64     FileForm.Tab1.Tab = 1
65   End If
66   Else
67     o = Shell(FileName, vbNormalFocus) ' Run any other shell program
68   End If
69 End If
70 End Sub
   ...
```

The code in Table 7.5 for the "RunExeFile" operation of a *MyFile* object is structured as follows. After checking the existence of file "FileName", its extension "FileType" and its mode (three "If" clauses starting in lines 02-04 and ending in lines 65-69), different blocks of code mostly for Mode 2 and 3 exe-files are sequentially executed to do the job. These **blocks** are:

1. Lines 05-06: Update I/O file titles and contents,
2. Lines 07-12: Go to FilePath directory, backup "FileName.txt" identical to "MyName.TXT" in Table 7.1 and print selected I/O file titles in "File-Name.txt",
3. Lines 13-18: Update contents of input files after the user edited them in the input window acc. to step 4 in Fig. 7.3.
4. Lines 20-40: Run the exe-file where the elapsed time is shown by a progress bar (line 30),
5. Lines 42-56: Restore "FileName.txt", update the output files and delete temporary files.
6. Lines 59-64: For Mode 1 exe-files: update the input file and run the program followed by a file dialog to select an output file (line 63), if no file was chosen before according to step 8 in Fig. 7.3,
7. Lines 67: Run any other exe-file that was selected in step 1 in Fig. 7.3.

Note that in the third block two input files are existing: the selected one and the one in the root directory of the exe-file called "FilePath + "\" + InpFTitle" (line 16) which is needed by the exe-file to start its processing.

What is left to explain in the sequence diagram (Fig. 7.5) of this application is the data plot of a data output file which has a variable number of columns. This is the main difference and challenge compared to the routine of the 2D dynamic data plotter in chapter 5 of this book (Table 5.4). It requires another operation called "PreparePlot" of a *MyFile* object whose code is listed in the following table.

Table 7.6. "PreparePlot" in file "MyFile.cls" in folder "Program04"

```
   ...
01 Public Sub PreparePlot()
02 If FileExists(OutFName) Then
03     Dim i As Integer, j As Integer, k As Integer
04     Dim s As Single, cols() As Single, st As String, status As Boolean
05     i = 0
06     Open OutFName For Input As #1 ' Open file for input
07        Line Input #1, st
08        Line Input #1, st
09        Line Input #1, st
10     If LCase(Left(st, 1)) <> "c" Then
11        Close #1 ' Close file if first 3 lines are comment lines
12        Open OutFName For Input As #1  ' Open file for input
13     End If
14        Do While Not EOF(1)       ' Loop until end of file
15           i = i + 1
```

```
16        Input #1, s
17        Loop
18      Close #1      ' Close file
19      If FileForm.Text2.Text = "" Then Call ReadOutput
20      Col = i / Max
21      Set SelectForm = New frmSelect
22      SelectForm.Text1 = Str(Max)
23      For i = 1 To Col ' Make column plot parameters visible
24          SelectForm.Label(i).Visible = True
25          SelectForm.XAxis(i).Visible = True
26          SelectForm.YAxis(i).Visible = True
27      Next i
28      SelectForm.Height = 2000
29      If Col > 3 Then SelectForm.Height = 2000 + 360 * (Col - 3)
30      SelectForm.XAxis(ColX).Value = 1
31      If ColYMax = 0 Then SelectForm.YAxis(Col).Value = 1
32      For i = 1 To ColYMax ' Get last selection of plot parameters
33          SelectForm.YAxis(ColY(i)).Value = 1
34      Next i
35      SelectForm.Left = FileForm.Left + 100
36      SelectForm.Top = FileForm.Top + 1400
37      SelectForm.Show vbModal, fGUI
38
39      For i = 0 To 10   ' Get selected plot parameters
40          If SelectForm.XAxis(i).Value = 1 Then ColX = i
41      Next i
42      ColYMax = 0
43      For i = 0 To 10
44          If SelectForm.YAxis(i).Value = 1 Then
45              ColYMax = ColYMax + 1
46              ColY(ColYMax) = i
47          End If
48      Next i
49      status = SelectForm.PlotStatus
50      Unload SelectForm
51       i = 0
52      ReDim cols(Col)
53      Set PointList = New Collection ' Create PointList
54      Open OutFName For Input As #1 ' Open file for input
55          Line Input #1, st
56          Line Input #1, st
57          Line Input #1, st
58      If LCase(Left(st, 1)) <> "c" Then
59          Close #1 ' Close file if first 3 lines are comment lines
60          Open OutFName For Input As #1  ' Open file for input
61      End If
```

```
62        Do While Not EOF(1)        ' Loop until end of file
63        i = i + 1
64        For j = 1 To Col
65            If Not EOF(1) Then Input #1, s
66            cols(j) = s
67        Next j
68
69        Dim poi As MyPoint        ' Create new object
70        Set poi = New MyPoint     '  of class MyPoint
71            poi.ID = Str(i)
72        If ColX = 0 Then
73            poi.X = i
74        Else
75            poi.X = cols(ColX)
76        End If
77        For j = 1 To ColYMax
78            k = ColY(j)
79            If j = 1 Then poi.Y1 = cols(k)
80            If j = 2 Then poi.Y2 = cols(k)
81            If j = 3 Then poi.Y3 = cols(k)
82            If j = 4 Then poi.Y4 = cols(k)
83            If j = 5 Then poi.Y5 = cols(k)
84            If j = 6 Then poi.Y6 = cols(k)
85            If j = 7 Then poi.Y7 = cols(k)
86            If j = 8 Then poi.Y8 = cols(k)
87            If j = 9 Then poi.Y9 = cols(k)
88            If j = 10 Then poi.Y10 = cols(k)
89        Next j
90        PointList.Add Item:=poi  ' Add poi to PointList
91        Set poi = Nothing
92        Loop
93     Close #1     ' Close file
94     If status = True Then Call PlotData(FileForm.Picture1)
95  End If
96  End Sub
       ...
```

The code in Table 7.6 for the "PreparePlote" operation of a *MyFile* object is structured as follows. After checking the existence of the output file "OutFName", different blocks of code are sequentially executed to prepare a plot. These **blocks** are:
1. Lines 03-05: Declaration of local variables,
2. Lines 06-20: Open output file and read its data to determine the maximum number of output data file lines called "Max" and the maximum number of columns named "Col" (line 20),

3. Lines 21-37:Prepare and show the *frmSelect* object named "SelectForm" where a user can select the wanted plot columns (Fig. 7.7),
4. Lines 39-50:The selected columns are stored in an array named ColY(ColYMax) where ColYMax can take up to a maximum of 10 columns,
5. Lines 51-93:The output file is opened again and the selected columns are placed into *MyPoint* objects named "poi" which are all added (line 90) to a *Collection* object named "PointList"(line 53),
6. Line 94: The "PlotData" operation is called, if "status" is true which is the case when a user clicked on the OK button of the *frmSelect* object in Fig. 7.7 (line 49).

Finally, the "PlotData" operation listed below is an extended version of the same operation in Table 5.4 for multiple Yi values, i = 1,2,...,10.

Table 7.7. Code for "PlotData" in file "MyFile.cls" in folder "Program04"

```
   ...
01 Public Sub PlotData(obj As Object)
02 If FileExists(OutFName) Then
03     If PointList.Count > 0 Then
04       obj.Cls  ' Clear drawing object
05
06     Dim i As Integer, xp As Single, yp As Single, yy As Single
07     Dim Color As Long, a As Single
08     Dim poi
09     Set poi = PointList.Item(1)      ' Start initialization
10       Par(1) = poi.X              ' Min for X
11       Par(2) = poi.Y1             ' Min for Y
12       Par(3) = poi.X              ' Max for X
13       Par(4) = poi.Y1             ' Max for Y
14
15     For Each poi In PointList
16       If Par(1) > poi.X Then Par(1) = poi.X
17       If Par(3) < poi.X Then Par(3) = poi.X
18
19       For i = 1 To ColYMax
20         yp = GetY(i, poi.Y1, poi.Y2, poi.Y3, poi.Y4, poi.Y5, _
21                 poi.Y6, poi.Y7, poi.Y8, poi.Y9, poi.Y10)
22         If Par(2) > yp Then Par(2) = yp
23         If Par(4) < yp Then Par(4) = yp
24       Next i
25     Next poi
26
27     If Par(1) - Par(3) <> 0 And Par(2) - Par(4) <> 0 Then
28       obj.Scale (Par(1), Par(4))-(Par(3), Par(2))  ' Scale drawing object
29       For i = 1 To 6                    ' Draw X axis in 5-er steps
```

```
30          xp = Par(1) + (i - 1) * (Par(3) - Par(1)) / 5
31          obj.Line (xp, Par(2))-(xp, Par(4)), RGB(220, 220, 220)
32          obj.CurrentX = xp
33          obj.CurrentY = Par(2) + 200 * (Par(4) – Par(2)) / obj.Height
34       If i = 6 Then obj.CurrentX = obj.CurrentX - 140 * (Par(3) - Par(1))/ _
35   obj.Width - 70 * Len(Format(xp, "###0.0")) * (Par(3) - Par(1)) / obj.Width
36          If i = 1 Then
37             obj.Print "X = " + Format(xp, "###0.0")
38          Else
39             obj.Print Format(xp, "###0.0")
40          End If
41          Yp = Par(2) + (i - 1) * (Par(4) - Par(2)) / 5  ' Draw Y axis in 5-er steps
42          Obj.Line (Par(1), yp)-(Par(3), yp), RGB(220, 220, 220)
43          Obj.CurrentX = Par(1)
44          Obj.CurrentY = yp
45   If i = 1 Then obj.CurrentY= obj.CurrentY + 400 *(Par(4) - Par(2))/ obj.Height
46          If i = 6 Then
47             Obj.Print "Y = " + Format(yp, "###0.0")
48          Else
49             Obj.Print Format(yp, "###0.0")
50          End If
51       Next i
52
53     For i = 1 To ColYMax
54       If ColYMax > 1 Then
55          a = (i - 1) / (ColYMax - 1) * ColYMax
56       Else
57          a = (i - 1)
58       End If
59          Color = GetColor(a, 0, ColYMax)
60
61     Set poi = PointList.Item(1) ' Draw lines from point to point
62          xp = poi.X
63          yp = GetY(i, poi.Y1, poi.Y2, poi.Y3, poi.Y4, poi.Y5, poi.Y6, poi.Y7, _
64   poi.Y8, poi.Y9, poi.Y10)
65     For Each poi In PointList
66          yy = GetY(i, poi.Y1, poi.Y2, poi.Y3, poi.Y4, poi.Y5, poi.Y6, poi.Y7, _
67   poi.Y8, poi.Y9, poi.Y10)
68          obj.Line (xp, yp)-(poi.X, yy), Color
69          xp = poi.X
70          yp = yy
71     Next poi
72     Next i
73
74       End If
75       End If
```

76	End If
77	End Sub
	...

The code in Table 7.7 for the "PlotData" operation of a *MyFile* object is structured as follows. After checking the existence of the output file "OutFName", the number of points in "PointList" and a minima/maxima difference greater than zero (three "If" clauses starting in lines 02-03, 27 and ending in lines 74-76), the transferred drawing object named "obj" (line 01) is cleared in line 04. Here, "obj" is a *PictureBox* object "Picture1" of an *mdiChild* object "FileForm" according to line 94 in Table 7.6.

The computational part of operation "PlotData" starts in line 06 with the declaration of all variables and the initialization of "Par()" needed to scale "obj" with the smallest and largest coordinate values (lines 10-13). After computing the "Par()" in lines 15-25, the drawing object "obj" is scaled in line 28 using the corresponding "Par()" values. Then a coordinate system is drawn and values of coordinate axis are written in 5-er steps (lines 29-51). Drawing of colored lines follows this from data point to data point for each Yi value (lines 53-72). These colors are computed in lines 54-59 so that they are distributed over the whole range of rainbow colors, using the same function "GetColor" already listed in Table 6.7.

7.4 Patterns used in this application

As mentioned in paragraphs 4.1 and 5.1, the **layers architectural pattern** with a user interface, a processor and a data manager is the fundamental three-tier architecture of all applications considered in part II of this book (Fig. 4.1 and 4.2). **Design patterns** used in this application are the same ones as in the first application "Program01":

Creational Patterns:

2. Builder: Separate the construction of a complex object from its representation so that the same construction process can create different representations.

3. Factory Method: Define an interface for creating an object, but let subclasses decide which class to instantiate. Factory Method lets a class defer instantiation to subclasses.

4. Prototype: Specify the kinds of objects to create using a prototypical instance, and create new objects by copying this prototype.

Behavioral Patterns:

16. Iterator: Provide a way to access the elements of an aggregate object sequentially without exposing its underlying representation.

19. Observer: Define a one-to-many dependency between objects so that when an object changes state, all its dependents are notified and updated automatically.

The **builder, factory method and prototype patterns** are used in this application when opening files and calling the factory method "CreateMyFile" (Table 4.3 and 4.4). Complex and different *MyFile* objects are created in the same way using the "ReadData" operation, which decides on which subclasses to instantiate (Table 7.3). The **iterator pattern** is being used with the *Collection* object "PointList" (Table 7.6 and 7.7) or "ChildList" (Table 7.4), and the **observer pattern** with the VB "WindowList" attribute of a *Menu* object (Fig 4.8), for example.

7.5 Testing the application

The interactive window shell program was tested with different exe- files in various ways in order to fulfill the requirements given in the beginning of paragraph 7.2. These tests can be categorized according to the application's three-tier architecture as follows:

- Testing the user interface,
- Testing the, processor and
- Testing the data manager.

Tests of the user interface and the data manager were not very difficult because of the gained insights when testing previous applications; for example, "Program02" which also does 2D data plots. In this application, the main difference and challenge was to incorporate multiple columns for a plot a user can select as wanted. The implementation was not too difficult as shown in the program code in Table 7.6.

Most of the test time of this application was spent on the processor, especially on operation "RunExeFile" of a *MyFile* object listed in Table 7.5. For example, line 20 in Table 7.5

20 | If FileExists(FilePath + "\Ok.end") Then Kill FilePath + "\Ok.end"

was firstly placed inside of the "Do While Loop" (lines 27-40) after calling the exe-file in line 26. The existence of file "Ok.end" indicates the end of the exe-file execution so that the progress bar can be stopped and the output file be automatically opened and displayed for a Mode 2 exe-file. But this caused run-time conflicts between functions "FileExists" and "Kill". Thus, it was decided to place line 20 before the "Do While Loop".

Another processing problem encountered was the display of an output file inside of an output window, which is a *TextBox* object named "Text2" with a maximum length of 32 kB as part of an *mdiChild* object named "FileForm". If the content of an output file is larger than this length, "Text2" is filled beyond its limit, causing the application into a "Not Responding" error. To prevent this to happen, the "ReadOutput" operation of a *MyFile* object should incorporate this limit and was changed as listed in the following Table 7.8.

Table 7.8. "ReadOutput" in file "MyFile.cls" in folder "Program04"

```
         ...
01 | Public Sub ReadOutput()
02 | If FileExists(OutFName) Then
03 |     Dim i As Integer, b As Boolean, s As String, nl As String
04 |     i = 0
05 |     b = False
06 |     nl = Chr(13) & Chr(10)
07 |     Open OutFName For Input As #1 ' Open file for input
08 |       FileForm.Text2.Text = ""
09 |       Do While Not EOF(1)      ' Loop until end of file
10 |       i = i + 1
11 |       Line Input #1, s
12 |       If i = 3 And LCase(Left(s, 1)) <> "c" Then i = i + 3
13 |       If LenB(FileForm.Text2.Text) < 32000 Then
14 |          FileForm.Text2.Text = FileForm.Text2.Text + s + nl
15 |       Else
16 |          If b = False Then
17 |             b = True
18 |             FileForm.Text2.Text = FileForm.Text2.Text + nl + _
19 |             "END OF DATA !!! The text box cannot show more data!" + nl _
20 |             + "Click on 'View/Output File' to see the whole output."
21 |          End If
22 |       End If
23 |       Loop
24 |     Close #1      ' Close file
25 |     Max = i - 3   ' Compute maximum number of output data file lines
26 | End If
27 | End Sub
         ...
```

The length of "Text2" in this operation is checked in line 13 by a VB routine called "LenB", which returns the length of a string in bytes. If this length is larger than 32000, the writing into the *TextBox* object "Text2" is finished with the message given in lines 19-20. The Boolean variable "b" is needed to guarantee that this message is only written once into "Text2".

In line 25, the maximum number of output data file lines called "Max" is computed assuming that the first three lines of an output file are comment lines with a "c" as first character like in the output window of Fig. 7.6. If this is not the case as checked in line 12, "Max" is incremented by a number of three.

CHAPTER 8: Image Digitizer with a Database in Visual Basic

8.1 Introduction to relational databases

This chapter describes a program to digitize points in a digital image that are stored in and managed by a database. This includes the adding, inserting and deleting of records as well as data queries. Databases allow centralized access to information in a way that is consistent, efficient and relatively easy to create and maintain. What follows is some basic knowledge of databases that is required to be able to understand the functionality of the designed program.

A **database** is a repository of information. While there are several different types of databases, this chapter deals only with relational databases, which are the most commonly used database type in the world today. A **relational database** (McManus 1999):
1. Stores data in tables, which are composed of rows (also called records) and columns (also called fields),
2. Enables you to retrieve (or query) subsets of data from tables, and
3. Enables you to connect (or join) tables together for the purpose of retrieving related records stored in different tables.

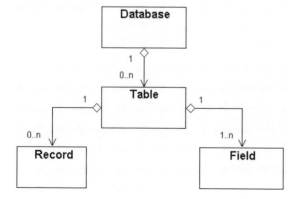

Fig. 8.1. Class diagram for a relational database

A class diagram for a relational database is shown in Fig. 8.1, where all classes have a whole-to-part association with each other. Classes *Record* and *Field* belong to class *Table*, which is a part of class *Database*.

A **database engine**, a software system that manages how data in a database is stored and retrieved, provides the basic functions of a database. The engine used in this program is called "Microsoft Jet". It is included in Visual Basic and can be simply used together with the Microsoft database program "Access", which was done in this chapter.

When **designing a database**, you must first decide on what information it should keep. You then design the database, creating tables composed of fields that define the types of data you want to store. After creating this database structure, the database can store and manage data in the form of records. It is not possible to add data to a database that has no table or field definitions because there is no place available to store it. Thus, the database design is crucial, particularly because it can be very difficult and time-consuming to change the design after it was implemented.

Table 8.1. Possible data types in VB databases

Data type	Description
Binary	Used for storing large data files like image or sound files
Boolean	A true or false value
Byte	A single-byte integer value between 0 and 255
Currency	A numeric field with special properties to store monetary values
Date/Time	An eight-byte date/time value between 01.01.100 and 31.12.9999
Double	An eight-byte double-precision numeric data type
GUID	A 128-byte number also called a globally unique identifier. It can be used to identify a record uniquely.
Integer	A two-byte whole number between –32,768 and 32,768
Long	A four-byte whole number between –2,147,483,648 and 2,147,483,648
Long Binary	A large-value field that can store binary data such as images or files
OLE Object	An OLE object embedded in a database can be up to 1 gigabyte
Memo	A large-value field that can store up to 65,353 characters. The length of this field needs not to be declared in advance.
Single	A four-byte single-precision numeric data type
Text	A fixed-length character field between 1 and 255 characters long
VarBinary	A variable binary data field

When designing tables, one of the steps in setting up the fields is to declare the type of each field, which enables the database engine to save and retrieve data much more efficiently. The available data types of the VB database Jet engine are listed in Table 8.1 (McManus 1999).

Another step when designing tables is to designate indexes. A database **index** is an attribute that can be assigned to a field to make it easier for the database engine to retrieve data based on information stored in that field. But there is a limit. Indexes make a database physically larger, so that too many indexes assigned to fields consume more memory and make the computer slower. In general, fields that are used most often in queries should be indexed. A **primary key** is a special type of index. A field with a primary key is used to uniquely identify a record. Thus, no two records in the same table may have the same value in its primary key field. Nor can a record be empty in that field. Every created table should have a primary key and should also be indexed on those fields that are queried the most.

After designing a database and storing data in tables, a way is needed to manipulate them. **Manipulating tables** includes entering, changing and retrieving data from tables. To do this, *Recordset* objects are used in combination with *Data* objects, both provided through the MFC library. The VB class *Data* provides access to data stored in databases using *Recordset* objects. It enables you to move from record to record and to display and manipulate data from records in data-aware controls, for example, a *DBGrid* object. This VB object allows assigning its "DataSource" property to the *Recordset* object of a *Data* object so that the corresponding table automatically fills the grid.

Most **data access operations** can be performed using *Data* objects without writing any code at all. Data-aware controls bound to *Data* objects automatically display data from one or more fields for the current record or, in some cases, for a set of records on either side of the current record. *Data* objects perform all operations on the current record. If the *Data* object is instructed to move to a different record, all data-aware controls automatically pass any changes to the *Data* object to be saved in the database. The *Data* object then moves to the requested record and passes back data from the current record to the data-aware control where it's displayed. *Data* objects automatically handle a number of operations including empty record sets, adding new records, editing and updating existing records as well as handling some common errors.

If a database contains two or more related tables, they can be linked by relationships. A **relationship** is a way of formally defining how two tables relate to each other. When you define a relationship, you tell the database engine which two fields in two related tables are joined. The two fields involved in a relationship are the primary key, introduced earlier in this chapter, and a foreign key. The foreign key is the key in the related table that stores a copy of the primary key in the main table.

The easiest **type of relationships** is a one-to-one relationship where one record in a table usually takes the place of one record in another table. More common is a one-to-many relationship in which each record in a table can have none, one or many records in a related table. A many-to-many relationship takes a one-to-many relationship a step further. A common example for this relationship is the relation between courses and students. Each student can have multiple courses and vice versa.

Normalization is a concept related to relationships. Basically, it dictates that a database should be consistent and work efficiently (McManus 1999). A database is inconsistent, for example, when data in one table does not match data entered in another table. Further, an inefficient database does not allow easily to isolate and extract wanted data. A fully normalized database, on the other hand, stores each piece of information in its own table and identifies it uniquely by its own primary key.

The **Structured Query Language** (SQL) is a standard way of manipulating and querying databases. It is available in various forms by almost all relational database systems including Microsoft "Access" applied in the application considered here. Generally, SQL is used for creating queries that extract data from databases, although a large subset of SQL commands perform other operations on databases, such as creating tables and fields. SQL commands can be divided into two categories (McManus 1999):
1. Data Definition Language (DDL) commands to create database components, and
2. Data Manipulation Language (DML) commands to retrieve records from databases.

A **query** is a database command that retrieves records from one or more tables. For example, the command "SELECT * FROM Points WHERE X > 100 ORDER BY Y" gets all records from table "Points" with an X coordinate greater than 100 in the order of the Y coordinates. The "SELECT" clause is at the core of every query to retrieve data. The "FROM" clause denotes a table as record source from which data is retrieved. The "WHERE" clause limits selected records according to a given a Boolean expression (here "X > 100"). The "ORDER BY" clause sorts all records in the order of a given field (here "Y"). Where is a query used? In principle, a query is used at all places where a table is accessed, such as where *Recordset* objects come into play.

8.2 Object-oriented analysis and design

The image digitizer with a database considered here should fulfill the following **requirements**. It should allow to:
1. Store points with their coordinates in a database attached to a digital image,
2. Insert, delete and edit any point record in the database, and
3. Search for digitized points in the database.

The class diagram of the image digitizer is based on the class diagram of the dynamic matrix processor in chapter 4 (Fig. 4.2) plus some other classes added to class *mdiChild* (Fig. 8.2). These classes are:

- class *Data1* to store and manage data points in a database,
- class *DBGrid1* to display a database table and edit its data, and
- class *frmQuery* for creating and running database queries.

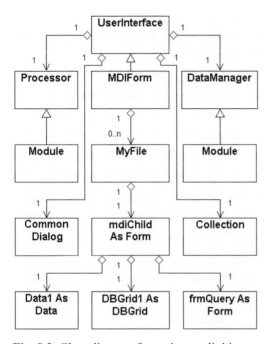

Fig. 8.2. Class diagram for an image digitizer

All three new classes have a whole-to-part association with class *mdiChild* and originate from a generalization-specialization association with classes *Data, DBGrid* and *Form* respectively, which belong all to the MFC library. The resulting class diagram is shown in Fig. 8.2.

A sequence diagram for the given requirements of this application is given in Fig. 8.3.

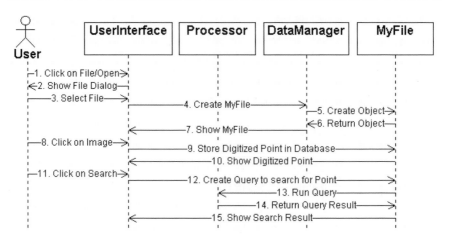

Fig. 8.3. Sequence diagram for an image digitizer

This diagram is very similar to the sequence diagram of the 2D dynamic data plotter (Fig. 5.3) in chapter 5. After selecting an image file, the corresponding *MyFile* object is created and displayed, which contains the image itself plus *Data1* and *DBGrid1* objects to store and manage digitized points. Digitizing of points can be done by either left mouse clicks inside of an image, or clicking on the "Add Record" button followed by entering the wanted point coordinates X and Y into the newly added record of the *DBGrid1* object. Searching for a digitized point starts by a click on "Search" in the program's menu bar, which produces an *frmQuery* object. This form holds two *TextBox* objects for entering X and Y coordinates of the point to be searched for. A click on the "Search" button of the *frmQuery* object then looks for this point in the database and encircles it, if the search was successful (Fig. 8.4). If not, a corresponding message is displayed.

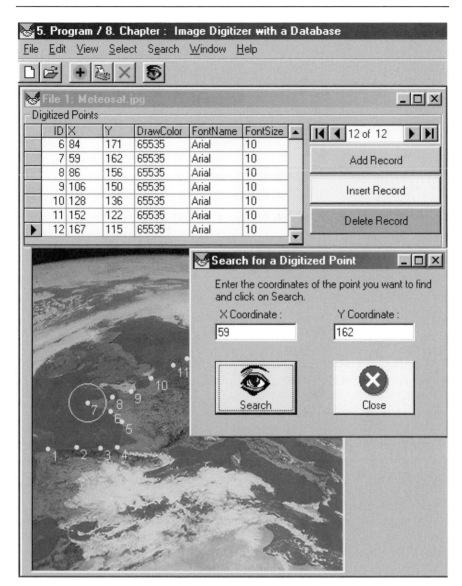

Fig. 8.4. Screen shot of the image digitizer with an *frmQuery* object

This figure shows three options in the designed application to add, insert or delete point records in a database. They are available in:
1. the menu bar of the user interface,
2. the icon bar of the user interface, or
3. in an *frmQuery* object through three command buttons named "Add/Insert/ Delete Record".

8.3 Implementation of the designed model

The coded architecture of the exe-file shell program can be easily visualized with the VB project explorer (McMonnies 2001) as shown in the figure below. This window is automatically displayed after opening the application's project file named "_Program05.vbp" in folder "Program05". All example programs including executable files and source code can be **downloaded** at http://de.geocities.com/bsttc2/book/SMOP.zip

Fig. 8.5. VB project explorer window for the image digitizer

As can be seen in this figure, "Program05" is very similar to "Program04" (Fig. 7.8) where file "mdiSelect.frm" was replaced by file "frmQuery.frm" and file "MyPoint.cls" was completely removed from this project.

How "Program05" works and how it was coded, can be more easily described by going through the application's sequence diagram (Fig. 8.3) and explaining step-by-step the corresponding program code. Because the beginning of "Program05" is almost identical to all other programs introduced before, a description of those parts can be omitted. The first important differences occur in operation "CreateMyFile" to set up a *MyFile* object with the following programming code (Table 8.2).

Table 8.2. "CreateMyFile" in file "DataManager.bas" in folder "Program05"

```
   | ...
01 | Public Function CreateMyFile(FTitle As String, FPath As String) As Object
02 |     Dim fi As MyFile      ' Create new object
03 |     Set fi = New MyFile    '  of class MyFile
04 |                            ' Assign input file attributes
```

```
05        fi.FileID = Str(ChildList.Count + 1)
06        fi.FileTitle = Ftitle
07        fi.FilePath = Fpath
08        fi.FileName = fi.FilePath + "\" + fi.FileTitle
09        If FileExists(fi.FileName) Then
10            fi.FileSize = FileLen(fi.FileName)
11        Else
12            fi.FileSize = 0
13        End If
14        fi.FileType = Right(fi.FileName, 3)
15        fi.FileStatus = True
16
17                    ' Assign child window attributes
18        fi.FileForm.FormID = fi.FileID
19        fi.FileForm.Caption = "File" + fi.FileID + ": " + fi.FileTitle
20                    ' Copy empty DB, if not existing
21        fi.FileForm.DBName = Left(fi.FileName, Len(fi.FileName) - 3) + "mdb"
22        If FileExists(fi.FileForm.DBName) = False Then _
23            FileCopy RootDir + "\System\ProtoDB.mdb", fi.FileForm.DBName
24
25        ChildList.Add Item:=fi  ' Add fi to ChildList
26
27        Call fi.ReadData       ' Read fi's data
28
29    Set CreateMyFile = fi
30
31    End Function
      ...
```

Lines 02-03 declare and create an object named "fi" of class *MyFile* including a child window, which is an *mdiChild* object named "FileForm", used to display the opened file and its data (step 7 in Fig. 8.3). The assignment of "fi" attributes happens in lines 05-23 depending on the given file parameters "FTitle" and "FPath" passed on to this function in line 01. These attributes of "fi" contain properties of the selected input file (lines 05-15) and attributes of "FileForm" (lines 21-23): a database name "DBName" (line 21) and an "Access" database file with the extension "mdb" copied from the program's system directory, if not existing (lines 22-23). When the attribute assignment is completed in Table 8.2, the new object "fi" is added to the "ChildList" in line 25 and its data is read in line 27. Finally, object "fi" is set equal to object "CreateMyFile" (line 29) and passed back to the calling object located in the "File/Open" operation of the user interface in file "GUI.frm". The "ReadData" operation of a *MyFile* object contains the following programming code (Table 8.3).

Table 8.3. Code for "ReadData" in file "MyFile.cls" in folder "Program05"

```
    ...
01  Public Sub ReadData()
02  If FileExists(FileName) Then
03      If LCase(FileType) = "bmp" Or LCase(FileType) = "gif" _
04      Or LCase(FileType) = "jpg" Or LCase(FileType) = "wmf" _
05      Or LCase(FileType) = "emf" Then
06        FileForm.Picture1.Picture = LoadPicture(FileName)
07        FileForm.Height = FileForm.Picture1.Height + 2620
08        FileForm.Picture1.ScaleMode = vbPixels
09        FileForm.Picture1.Visible = True
10
11        FileForm.Data1.DatabaseName = FileForm.DBName ' Connect to DB
12        FileForm.Data1.RecordSource = "Coordinates"
13        FileForm.Data1.Refresh
14        FileForm.DBGrid1.Columns(0).Width = 400
15        FileForm.DBGrid1.Columns(1).Width = 600
16        FileForm.DBGrid1.Columns(2).Width = 600
17        FileForm.DBGrid1.Columns(3).Width = 900
18        FileForm.DBGrid1.Columns(4).Width = 1280
19        FileForm.DBGrid1.Columns(5).Width = 740
20
21        FileForm.DrawColor = vbYellow
22        FileForm.DrawFontName = "Arial"
23        FileForm.DrawFontSize = "10"
24        Call FileForm.DrawPoints
25
26      End If
27  End If
28  End Sub
    ...
```

After checking the existence of file "FileName" of a *MyFile* object, a second "If" clause is used to check its file type in lines 03-05. If it is an image file, further attributes of "FileForm" are assigned in lines 21-23 and finally its operation "DrawPoints" is called in order to display digitized points inside of the image stored in the database. Before this can happen, the image is loaded into a *Picture-Box* object named "Picture1" belonging to the child window object "FileForm" (line 06). Its height is then resized according to the height of "Picture1" (line 07), whose scale mode is set to pixels (line 08) and made visible in line 09.

In lines 11-13, the attributes of *Data* and *DBGrid* objects named "Data1" and "DBGrid1" are assigned. "Data1" is connected to the database in line 11. Its record source is the table called "Coordinates" (line 12) and "Data1" together with "DBGrid1" are updated in line 13. Then the widths of all existing columns in

"DBGrid1" are selected (lines 14-19). Finally, drawing color, font name and size are assigned before existing points in table "Coordinates" are plotted in the image. The programming code for this operation is listed in Table 8.4.

Table 8.4. Code for "DrawPoints" in file "mdiChild.frm" in folder "Program05"

```
   ...
01 Public Sub DrawPoints()
02      Picture1.Cls
03 If Data1.Recordset.RecordCount > 0 Then
04      Dim i As Long, sx As String, sy As String, _
05          s1 As String, s2 As String, s3 As String
06      Data1.Recordset.MoveFirst
07      While Val(Format((Data1.Recordset.RecordCount * _
08      (Data1.Recordset.PercentPosition * 0.01)), "###0")) _
09      < Data1.Recordset.RecordCount
10          i = Data1.Recordset.Fields("ID")
11          sx = Data1.Recordset.Fields("X")
12          sy = Data1.Recordset.Fields("Y")
13          s1 = Data1.Recordset.Fields("DrawColor")
14          s2 = Data1.Recordset.Fields("FontName")
15          s3 = Data1.Recordset.Fields("FontSize")
16
17          If sx <> "?" And sy <> "?" Then
18              Picture1.FillColor = s1
19              Picture1.Circle (Val(sx), Val(sy)), 2, s1
20              Picture1.ForeColor = s1
21              Picture1.FontName = s2
22              Picture1.FontSize = Val(s3)
23              Picture1.CurrentX = Val(sx) + 1
24              Picture1.CurrentY = Val(sy)
25              Picture1.Print Str(i)
26          End If
27
28          Data1.Recordset.MoveNext
29      Wend
30 End If
31 End Sub
   ...
```

In the beginning of this drawing operation, "Picture1" is cleared in line 02 and an "If" clause checks the existence of any points in the data table (line 03). If the record count is greater zero, the code block in lines 04-29 is executed, containing a "While-Wend" loop (lines 07-29), which moves from the first to the last record using the "MoveNext" operation of the *Recordset* object of a *Data* object (line 28). The data of each record is read in lines 10-15 and displayed in lines 18-25, if the X and Y coordinates of a point are unequal to the question mark character (line 17; see lines 07-08 in Table 8.5). This is the case for a newly added point after clicking on the "Add Record" command button in Fig. 8.4. The code is shown in the table below.

Table 8.5. "Command1_Click" in file "mdiChild.frm" in folder "Program05"

```
    ...
01  Public Sub Command1_Click()
02      Dim n As Long
03      n = LargestID("Coordinates") + 1
04      With Data1.Recordset
05          .AddNew
06          !ID = n
07          !X = "?"
08          !Y = "?"
09          !DrawColor = DrawColor
10          !FontName = DrawFontName
11          !FontSize = DrawFontSize
12          .Update
13      End With
14      Data1.Recordset.MoveLast
15      DBGrid1.Col = 1
16      DBGrid1.SetFocus
17  End Sub
    ...
```

After computing the largest ID in table "Coordinates" via a function named "LargestID" (line 03; see Table 8.6), a new record to the Recordset object is added (line 05) inside of a "With" clause in lines 04-13 where the X and Y coordinates are left open to be entered later by keyboard input (lines 07-08).

Table 8.6. Code for "LargestID" in file "mdiChild.frm" in folder "Program05"

```
    ...
01  Private Function LargestID(TableName As String) As Long
02      Dim dbTmp As Database
03      Dim qdfTmp As QueryDef
04      Dim rsTmp As Recordset
05      Set dbTmp = OpenDatabase(DBName)
06      With dbTmp
07          Set qdfTmp = .CreateQueryDef("", _
08          "SELECT * FROM " + TableName + " ORDER BY ID")
09          Set rsTmp = qdfTmp.OpenRecordset()
10          If rsTmp.RecordCount > 0 Then
11              rsTmp.MoveLast
12              LargestID = rsTmp.Fields("ID")
```

13	Else
14	LargestID = 0
15	End If
16	End With
17	End Function
	...

This function returns the largest ID in a table to a caller using a SQL string with an "ORDER BY" clause to ensure that really the largest ID is depicted (line 08). Before a temporary database named "dbTmp" is declared in line 02 and opened in line 05. Its *Recordset* object "rsTmp" is filled with a query definition "qdfTmp" using the VB "OpenRecordset" procedure (line 09). The query definition itself is assigned to an SQL string via the VB "CreateQueryDef" procedure (lines 07-08).

As mentioned before, the other option to digitize points it to just click on the image at the wanted position, which represents steps 8-10 in the sequence diagram in Fig. 8.3. The programming code for this functionality is given in Table 8.7.

Table 8.7. Code for "Picture1_MouseDown" in file "mdiChild.frm" in folder "Program05"

```
01 | Private Sub Picture1_MouseDown(Button As Integer, Shift As Integer, _
02 |                                 X As Single, Y As Single)
03 | If Button = vbLeftButton Then
04 |      Dim s As String
05 |      Call Command1_Click
06 |      With Data1.Recordset
07 |         .Edit
08 |         s = Str(!ID)
09 |         !X = Str(X)
10 |         !Y = Str(Y)
11 |         !DrawColor = DrawColor
12 |         !FontName = DrawFontName
13 |         !FontSize = DrawFontSize
14 |         .Update
15 |      End With
16 |      DBGrid1.Col = 0
17 |
18 |      Picture1.FillColor = DrawColor
19 |      Picture1.Circle (X, Y), 2, DrawColor
20 |      Picture1.ForeColor = DrawColor
21 |      Picture1.FontName = DrawFontName
22 |      Picture1.FontSize = Val(DrawFontSize)
23 |      Picture1.CurrentX = X + 1
24 |      Picture1.CurrentY = Y
```

```
25          Picture1.Print s
26   End If
27   End Sub
     ...
```

After checking the left mouse button was clicked in line 03, the operation "Command1_Click" in Table 8.5 is called (line 05) and then this newly created record edited in lines 06-16. The only difference to Table 8.5 is now that the X and Y coordinates from the mouse click (line 02) are immediately stored in the database in lines 09-10 plus line 14. A circle marks the digitized point (line 19) that is filled with the selected drawing color (line 18). A label is printed (line 25) at its position (lines 23-24) with the selected font name and size. The code for these selections is listed in the table below. All three are available after clicking on "Select" in the program's menu bar.

Table 8.8. Code for color, font name and size selections in file "fGUI.frm" in folder "Program05"

```
     ...
01   Private Sub mnuSelectDrawColor_Click()
02   If ChildList.Count > 0 Then
03      With FileDialog
04         .color = 0
05         .ShowColor
06         If .color > 0 Then
07            Dim obj
08            For Each obj In ChildList   ' Find child in list
09               If Val(Screen.ActiveForm.FormID) = Val(obj.FileID) Then
10                  obj.FileForm.DrawColor = .color
11               End If
12            Next obj
13         End If
14      End With
15   End If
16   End Sub
17
18   Private Sub mnuSelectFont_Click()
19   If ChildList.Count > 0 Then
20      With FileDialog
21         .FontName = ""
22         .FontSize = 0
23         .Flags = cdlCFScreenFonts
24         .ShowFont
25         If .FontName <> "" And .FontSize > 0 Then
26            Dim obj
27            For Each obj In ChildList   ' Find child in list
```

28	If Val(Screen.ActiveForm.FormID) = Val(obj.FileID) Then
29	obj.FileForm.DrawFontName = .FontName
30	obj.FileForm.DrawFontSize = .FontSize
31	End If
32	Next obj
33	End If
34	End With
35	End If
36	End Sub
	...

There is nothing unusual in Table 8.8 apart from line 23, which has to be included when a *CommonDialog* object like "FileDialog" calls the VB "ShowFont" procedure in order to get all screen fonts into the dialog. Note that every *mdiChild* object "FileForm" can have different selections that are only available after an image was opened and "FileForm" was created. This is checked by the "If" clauses in lines 02 and 19.

When the "Search" option in the program's menu bar is selected in Fig. 8.4, which represents steps 11-15 in the sequence diagram in Fig. 8.3, the *frmQuery* object is displayed so that X and Y coordinates can be entered into two *TextBox* objects of the point one wants to find. Then a press on the "Seach" command button executes the following code in Table 8.9.

Table 8.9. Code for "Command1_Click" in file "frmQuery.frm" in folder "Program05"

	...
01	Private Sub Command1_Click()
02	Call RunQuery(fForm, Text1.Text, Text2.Text)
03	End Sub
	...

The operation "RunQuery" (line 02) contains three arguments. The first one, "fForm", takes the *mdiChild* object "FileForm". The other two ones hold the X- and Y coordinates. The programming code of this operation is listed in Table 8.10.

Table 8.10. Code for "RunQuery"in file "Processor.bas" in folder "Program05"

	...
01	Public Sub RunQuery(frm As Form, xText As String, yText As String)
02	Dim dbTmp As Database
03	Dim qdfTmp As QueryDef
04	Dim rsTmp As Recordset
05	Set dbTmp = OpenDatabase(frm.DBName)
06	With dbTmp
07	Set qdfTmp = .CreateQueryDef("", _

```
08 │        "SELECT * FROM Coordinates WHERE X = '" + xText _
09 │                    + "' AND Y = '" + yText + "'")
10 │        Set rsTmp = qdfTmp.OpenRecordset()
11 │        If rsTmp.RecordCount > 0 Then
12 │            Call frm.DrawPoints
13 │            frm.Picture1.FillStyle = 1
14 │            frm.Picture1.Circle (Val(xText), Val(yText)), 20, frm.DrawColor
15 │            frm.Picture1.FillStyle = 0
16 │        Else
17 │        MsgBox "The entered point with coordinates X = '" + xText + "' and _
18 │                    Y = '" + yText + "' could not be found in the database", _
19 │                    vbInformation, fGUI.Caption
20 │            End If
21 │        End With
22 │ End Sub
   │    ...
```

There is no principle difference between the code in Table 8.6 and this code apart from the SQL string in lines 08-09 and the circle around a selected point that is drawn, if the search was successful (lines 12-15; see Fig. 8.4). Otherwise, a message box is displayed with the message that the entered point coordinates could not be found in the database (lines 17-19).

8.4 Patterns used in this application

As mentioned in paragraphs 4.1 and 5.1, the **layers architectural pattern** with a user interface, a processor and a data manager is the fundamental three-tier architecture of all applications considered in part II of this book (Fig. 4.1 and 4.2). **Design patterns** used in this application are the same ones as in the first application "Program01":

Creational Patterns:

2. Builder: Separate the construction of a complex object from its representation so that the same construction process can create different representations.

3. Factory Method: Define an interface for creating an object, but let subclasses decide which class to instantiate. Factory Method lets a class defer instantiation to subclasses.

4. Prototype: Specify the kinds of objects to create using a prototypical instance, and create new objects by copying this prototype.

Behavioral Patterns:

16. Iterator: Provide a way to access the elements of an aggregate object sequentially without exposing its underlying representation.

19. Observer: Define a one-to-many dependency between objects so that when an object changes state, all its dependents are notified and updated automatically.

The **builder, factory method and prototype patterns** are used in this application when opening files and calling the factory method "CreateMyFile" (Table 8.2). Complex and different *MyFile* objects are created in the same way using the "ReadData" operation, which decides on which subclasses to instantiate (Table 8.3). The **iterator pattern** is being used with the *Collection* object "ChildList" (Table 8.2), and the **observer pattern** with the VB "WindowList" attribute of a *Menu* object (Fig 4.8), for example.

8.5 Testing the application

The image digitizer was tested with different image files in various ways in order to fulfill the requirements given in the beginning of paragraph 8.2. These tests can be categorized according to the application's three-tier architecture as follows:
- Testing the user interface,
- Testing the, processor and
- Testing the data manager.

Tests of the processor interface and the data manager were not very difficult because of the gained insights when testing previous applications, for example, "Program02" which also does data point plotting. In this application, the main difference and challenge was to incorporate a database to store, manage and search for digitized points. For this, the main building blocks are *Data, Recordset* and *DBGrid* objects.

Why was the *Collection* object "PointList" omitted in this program? The answer is that in previous applications, "PointList" was used to manage and plot data points. Here, "PointList" was not included because *Data* and *Recordset* objects handle its functionality now.

Regarding the SQL, there is of course a lot to learn for a beginner. Depending on the database engine and SQL dialect used, commands may differ slightly from each other. When designing queries, it is always important to keep in mind the data type of the query variables (Table 8.1). They should be the same in a single Boolean expression. This can cause conflicts for variables that are used with mixed data types like a number and a text string so that a data type conversion is required at some place or the other.

In such a situation, the author recommends to select the text data type as base type because it is more flexible regarding all the different factors involved. For this, an example can be found in lines 09-10 in Table 8.7 where the X and Y coordinates of a digitized point are converted from type "Single" to type "Text" with the VB "Str" procedure. Note that in queries, text variables have to be enclosed by single quotation signs (') as part of a SQL string, as was done in lines 08-09 in Table 8.10.

Another problem encountered regarding SQL, *Data* and *Recordset* objects was the question how to delete and then add or insert records in a table so that identical indexes are avoided. The solution was already presented in Table 8.6 where the operation "LargestID" returns the largest ID in a table to a caller using a SQL string with an "ORDER BY" clause to ensure that really the largest ID is depicted (line 08 in Table 8.6).

In lines 11-12 in Table 8.3, the *Data* object "Data1" is connected to a database and its record source to a table called "Coordinates". This can be done with a database that is not password-protected. But if a database has a **password**, the following alternative exists instead of lines 11-12 in Table 8.3 (McManus 1999).

Table 8.11. Alternative code for a password-protected database in operation "ReadData" in file "MyFile.cls" in folder "Program05" (the password is "pw")

```
      ...
11       Dim dbSpace As Workspace
12       Dim dbFile As Database
13       Set dbSpace = CreateWorkspace("", "admin", "", dbUseJet)
14       Set dbFile = dbSpace.OpenDatabase("", True, False, _
15             ";DATABASE=" & FileForm.DBName & ";UID=;PWD=pw")
16       FileForm.Data1.DatabaseName = FileForm.DBName
17       FileForm.Data1.Connect = dbFile.Connect
18       FileForm.Data1.RecordSource = "Coordinates"
      ...
```

Here, the database connection is established through a VB *Workspace* object named "dbSpace" and a VB *Database* object named "dbFile" declared in lines 11-12. A VB *Workspace* object defines a named session, the time between connecting and disconnecting to a database. It contains open databases and provides mechanisms for simultaneous transactions and, for the "Microsoft Jet" database engine used here, secures workgroup support. The *Workspace* object "dbSpace" is created in line 13 so that in lines 14-15, the *Database* object "dbFile" can be opened with the VB "OpenDatabase" routine of the *Workspace* object "dbSpace". This routine has the following syntax ("[]" means optional):

Set database = workspace.OpenDatabase (dbname, [options], [read-only], [connect]) where "dbname" is the only required argument – the name of the database to be opened. If this *String* object is set to the null string as in line 14 in Table 8.11, the connection parameters are provided through the optional "connect" *String* object, containing the database name ("DATABASE"), user ID ("UID") and a password ("PWD"), which is set to "pw" in line 15. The other two optional parameters of "OpenDatabase" are "options" and "read-only". If "options" is set to "True", the database is opened in exclusive mode and cannot be opened elsewhere. If "read-only" is set to "True", no changes can be made to the database. Therefore, it is set to "False" in line 14 because changes are allowed in this application. Finally, the *Data* object "Data1" is connected to "dbFile" in line 16.

Regarding the font name and size selections in Table 8.8, it is important that when using the VB "ShowFont" procedure of a *CommonDialog* object, the "Flags" attribute has to be set to one of the three following VB constants:
1. "cdlCFBoth",
2. "cdlCFPrinterFonts", or
3. "cdlCFScreenFonts",

as was done in line 23 of Table 8.8. If "Flags" is not set, a run-time error occurs with the message "No fonts exist".

CHAPTER 9: City Map Java Applet with a Road Finder

9.1 Overview of Java

Java is a true object-oriented language that incorporates some of the most power-ful new concepts in computer language design introduced in the last two decades. Java was designed to make programmers more productive by hiding much of the complexity that is visible in other languages. While Java strictly enforces many rules at compile time, it also includes features to allow extremely dynamic code to be written. Java includes native support for multi-threaded development, makes use of modern exception handling, and applies sophisticated garbage collection techniques to take care of memory management automatically (http://java.sun.com).

Java is a compiled language, but the object code is not machine specific. Any computer with a Java Virtual Machine can run the same compiled Java code. A comprehensive collection of standard Java classes gives the Java developer a head start in developing database, graphical or network applications. All of these fea-tures are important, but what is truly unique about Java is the level of industry ac-ceptance that it has received. No other language has seen wide spread support and acceptance this rapidly in the history of the computing industry.

Java programs can take many forms. The most commonly discussed form of a Java program is the applet. An applet is Java code that is automatically downloa-ded from the Web and runs within a Web browser on the user's desktop. This install-free approach to distributing and running software is extremely attractive for some markets, and Java's built-in security provides assurances that the applet cannot misbehave within its limited environment. Java applets are being used to collect data, display graphical summaries, and provide interactive entertainment over the Web.

Java can be used to develop full-fledged applications as well. These applica-tions are installed on an individual system or network in a more traditional man-ner, and are free of the security restrictions placed on applets, giving them the ability to perform almost any task that programs written in other languages can perform. Java has outstanding multi-threading and network connectivity libraries, and applications written entirely in Java can run on a variety of systems without needing recompilation. Java developers are able to take advantage of this highly productive language to develop new kinds of software, including distributed applications and collaborative tools.

As already mentioned, Java is a compiled language but rather than producing machine code for a specific processor, the Java compiler generates Java byte codes, which are interpreted by a Java Virtual Machine (JVM). Providing a computer or workstation has an implementation of the JVM, any Java program can be run on that machine. This allows a Java program to be compiled once, stored on a server, transferred across the network to a client machine and then executed without worrying about what kind of client machine it is running on. The program code is said to be architecture neutral, as its representation is independent of both the physical architecture of a machine and the operating system it is running. This is a very important issue as software developers no longer have to worry about targeting a particular operating system and can rely on their program running on any machine with an implemented JVM.

Other kinds of Java programs are appearing regularly, including server technologies and operating systems. Java is still maturing, and while it is currently being used most often to solve problems that more mature technologies do not address, it is expected to achieve mainstream acceptance for other tasks as it evolves.

9.2 Introduction to Java applets

Java applets are a new way of distributing code to the desktop. Unlike a traditional application, which must be installed before use, an applet can be downloaded and executed automatically when a user visits a Web page with a standard Java-enabled Web browser. This allows designers of an Internet site to provide interactive experiences. The same technology can be used on corporate intranets to deploy custom applications to desktops while reducing the administrative burden required keeping these desktops up-to-date.

Applets use the strengths of the Java language in a number of ways. The Web browser needs to be able to dynamically load new classes in order to execute applets, and Java is well suited to this task. The code must work on a variety of platforms, so users in multiple operating systems can take advantage of the downloaded code. Java's platform-independent byte code solves this problem neatly. The code that is downloaded must not be able to adversely affect the user's computer. Ideally, it should be impossible to write a virus in Java, and the design of the language and class libraries restricts applets in a number of critical ways to achieve this goal.

Java applications and applets share the same language, the same standard packages, and differ only slightly in their entry points. The choice between the two types of Java programs is usually answered by a single **question** — "Do you want code automatically delivered from a Web server into a browser environment?" If so, an applet is the way to go. If not, an application is the right stand-alone solution. Simple, isn't it? Unfortunately, the decision isn't as simple as it seems. There are some pretty significant restrictions placed on applets by default, and some capabilities that are available to applets aren't supported for applications. Worse yet, sometimes the same code needs to he delivered both as an applet and an application, and consequently, both sets of restrictions become an issue. Some of the major differences between applications and applets are the applet security restrictions described next.

Applets are typically subject to a wide range of **security restrictions** designed to prevent them from performing undesirable actions when downloaded from a Web server. The following restrictions are enforced by default:
1. Applets cannot access the local file system to read and write, nor can they determine the existence or status of a file.
2. Networking is severely restricted, effectively preventing networking connections to any machine other than the Web server that originally supplied the applet.
3. Clipboard and printing support are inaccessible.
4. Frame and Dialog instances created by applets include a visible warning that the window has been created by not trusted code.
5. Java facilities for starting applications, inquiring about the system and user, and stopping the Java interpreter, for example, are not available.

Attempting to perform any of these or other security-violating actions will result in a security exception error. Typically, it is better to design code to avoid these situations than to rely on extensive exception handling to deal with them.

An applet **consists of two parts**: a body of Java code that describes the actual behavior of the applet, and an HTML file that describes a Web page the applet should be displayed on (Deitel & Deitel 1997).

The Java portion of an applet is a subclass of *java.applet.Applet.* This subclass defines the behavior unique to that particular applet through methods and instance variables just like any other Java class. When the applet is needed, a Web browser will download the class, create an instance of the class, and to user input. If, in turn, the applet calls upon additional classes, images, or sounds, they will also be downloaded from the Web server. For added efficiency, multiple classes and resources, such as images and sounds, may he bundled together in a single download in JAR (Java Archive) file format

The HTML file provides a context for the applet and information about the location, size, and optional parameters passed to the applet. This information is described using an <APPLET> tag and special attributes. More than one <APPLET> tag can be used in a single HTML file, allowing for any number of applets on a single page.

An applet is always an instance of a class that extends *java.applet.Applet* and adds its own unique behavior by overriding some of the *Applet's* methods. The *Applet* class is a container, inheriting from *java.awt.Panel,* and like all containers, is itself a customizable component. Therefore, it may contain any number of other components that can be arranged manually or with any of the standard layout managers. This technique is well suited to developing user interfaces, using the applet itself solely as a container for more interesting components, for example, the ones from Java's Abstract Windowing Toolkit (AWT).

An applet can be designed without overriding a single *Applet* method, but typically one or more of these methods will be overridden so that the applet can take advantage of information about the context it is displayed within.

The Web browser notifies an applet that it has been created and configured by invoking its "init()" method. Applets should typically override this method to perform initialization actions rather than placing initialization code in their constructor. An applet's "destroy()" method is called just before the applet is discarded. This method is rarely overridden since garbage collection will typically take care of the necessary cleanup.

When the applet is actually displayed, its "start()" method is invoked so an applet can override this method to perform any actions related to being displayed. This is typically when an applet would begin any animation or periodic task that is important while it is visible. If the "start()" method is overridden, the corresponding "stop()" method should also be overridden to cease this activity when the browser moves on to another page. A single applet may be started and stopped several times before being discarded.

The methods "getAppletInfo()" and "getParameterInfo()" provide additional information about the applet, but do not directly affect its behavior. The method "getAppletInfo()" returns a string suitable for display that describes the applet and should always he overridden to provide meaningful information. The method "getParameterInfo()" describes the parameters the applet accepts, and should be overridden for any applet that accepts parameters. The most frequently overridden method from *java.applet.Applet* is the "paint()" method, which describes the applet's appearance. It paints any components contained inside the applet on the screen. Any applet can further customize its appearance by overriding this method. The sample applet presented next defines an applet that uses "paint()" to create its own appearance.

9.3 Creating a simple "Hello World" applet

In this paragraph, a simple applet is introduced consisting of some Java code that determines the applet's behavior, and an HTML file for displaying the applet inside of a Web page. The Java code is given in the following table.

Table 9.1. Code of file "SimpleApplet.java" in folder "Program06"

```
       ...
01 | import java.awt.*;
02 | public class SimpleApplet extends java.applet.Applet {
03 |     public void paint(Graphics g) {
04 |         g.setColor(Color.blue);
05 |         g.setFont(new Font("Arial", Font.BOLD, 24));
06 |         g.drawString("Hello World!", 40, 60);
07 |         g.drawRect(20, 20, 180, 70);
08 |     }
09 | }
       ...
```

A look at the code in Table 9.1 reveals that *SimpleApplet* draws the string 'Hello World!' in blue using a 24-point bold Arial font, and then draws a blue rectangle around the text as shown in Fig. 9.1. The content of the HTML file for this application is listed in Table 9.2.

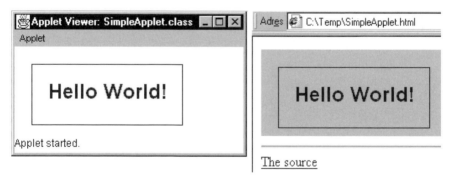

Fig. 9.1. Screen shot of the simple applet in applet viewer and Web browser

The code in Table 9.2 shows the simplest form of an <APPLET> tag inside of an HTML file. It has to contain the required attributes code, width and height in pixel, which reserve a rectangular area with the given width and height (here 280x100 pixel) for the applet inside of the HTML document. Line 08 holds a reference to the source in Table 9.1, which can be opened after clicking on the text string "The source" in the HTML document.

Table 9.2. Code of file "SimpleApplet.html" in folder "Program06"

```
      ...
01  <title>SimpleApplet</title>
02  <APPLET CODE = "SimpleApplet.class"
03              WIDTH = 280
04              HEIGHT = 100
05  >
06  </APPLET >
07  <hr>
08  <a href="SimpleApplet.java">The source</a>
09      ...
```

Web browsers attempt to display applets wherever the <APPLET> tag is encountered in the HTML that describes a Web page. The <APPLET> tag reserves a rectangular area within the page for the applet's use, and tells the browser how to load and start the applet class. The <APPLET> tag's syntax is given in Table 9.3.

Table 9.3. Syntax of the <APPLET> tag [parts in square brackets are optional]

```
      ...
01      <applet
02      CODE = fully qualified path CLASS
03      WIDTH = width in pixels
04      HEIGHT = height in pixels
05      [ OBJECT = serialized applet url ]
06      [ ARCHIVE = jar/zip file list ]
07      [ CODEBASE = class path root url ]
08      [ ALT = no applet alternate text ]
09      [ NAME = applet name ]
10      [ ALIGN = alignment ]
11      [ VSPACE = vertical spacing in pixels ]
12      [ HSPACE = horizontal spacing in pixels ]
13      >
14      [ Alternate text ]
15      </APPLET>
      ...
```

The ALT attribute is only displayed by Java-aware browsers that have applets disabled, while the "Alternate text" contained between the <APPLET> and </APPLET> tags is displayed by browsers that are not Java-aware.

The required CODE attribute specifies the class to be loaded as an applet. The value specified must include the ".class" extension for the class file stored on the Web server, and use the Java dot notation for specifying the package name for the class. The path on the Web server from which code will be requested is adapted from this description. For example, if line 02 in Table 9.2 would read

02 | <APPLET CODE = "examples.program06.SimpleApplet.class"

then the class described by the relative URL (Uniform Resource Locator) "examples/program06/SimpleApplet.class" would be requested from the Web server and loaded as an applet.

The optional CODEBASE attribute allows changing the location from where Java code is requested by a Web browser. Normally, the browser will request an applet's class files from an URL relative to the HTML file that contains the <APPLET> tag. If the "SimpleApplet.class" from the previous example is linked to e.g. an URL "http://www.abc.net/applet/SimpleApplet.html", then the browser tries to load "http://www.abc.net/applet/examples/program06/SimpleApplet.class" to find the class definition. But if, for example, the following code base is used

07 | CODEBASE = "/book"

the browser would search for "http://www.abc.net/applet/book/examples/ program06/SimpleApplet.class".

The ARCHIVE attribute also changes the way the Web browser requests Java code from a Web server. When using the default loading mechanism, classes are loaded one at a time from the Web server, which can be extremely inefficient. The ARCHIVE attribute specifies a compressed Java archive (JAR) or ZIP file on the server that will be searched for the class in question. For example, the following code

06 | archive = "/book/SimpleApplet.zip"

makes the browser first download file "http://www.abc.net/applet/ book/SimpleApplet.zip" and then look inside it for "examples/program06/ SimpleApplet.class". When searching for additional classes, the browser will search the archive that has already been downloaded and placed in the browser's cache, thus reducing the time required to load and run an applet from a remote Web server.

To **run or debug an applet**, the applet viewer as a part of the Java Developer's Kit (JDK) from Sun Microsystems Inc. (http://java.sun.com) can be used. This minimal applet host environment runs inside its own Java Virtual Machine, and creates a simple frame with a status line for display purposes as can be seen on the left-hand side of Fig. 9.1. The applet viewer does enforce applet security and provides a realistic test environment for most purposes, but it cannot be used as a guarantee that an applet will behave properly in other Web browsers.

A major disadvantage of the applet viewer is that it provides extremely limited feedback about error conditions and sources. A better alternative is to view the execution log of an applet in the console window (Fig. 9.2) to be able to fix an error. Besides, an applet that works correctly in the applet viewer may fail to work within a Web browser for a variety of reasons like an access error as shown in Fig. 9.2.

Fig. 9.2. Console window displaying an access violation

9.4 Creating a city map applet with a road finder

The objective of this paragraph is to give an overview of how to create an applet for viewing a city map and a search routine to find roads in this map. Input data provided for this task is a city map in form of one or more digital image files and a database containing road names with their corresponding map location.

This applet gives at least three challenges to the model builder.
1. The city map is generally much larger than the display area of the applet so that some type of image scrolling is required.
2. Dense parts of a city map make a zooming function necessary.
3. A road found through the search routine needs to be moved to and marked so that a user can easily detect it.

The best solution for the latter challenge was thought to be a blinking text label with an "i" for information.

So far, every application was based on a 3-tier architecture consisting of user interface, processor and data manager, which were placed in different modules. Because applets are very small Java programs by nature and the application considered here is not very complex, the 3-tier approach was abandoned. Instead, all needed classes and objects are placed in one single module and Java file called "Program06.java", which has the same basic structure as "SimpleApplet.java" in Table 9.2. After importing all needed class libraries in the beginning, the applet class *Program06* is defined as an extension of class *java.applet.Applet* through a generalization-specialization association. The class body of *Program06* consists of declarations of attributes and operations as well as other objects including their attributes and operations.

Before starting with the object-oriented analysis and design of this application, the input data is described in more detail. The city map is given through a file "CityMap.GIF" that standard graphic classes can easily open like the one used here, which is *borland.jbcl.control.TransparentImage* from Borland Inc. (http://www.borland.com). "JBCL" stands for "Java Beans Component Library". The MS Access file "CityMap.MDB" is the database with the road names and their location. It contains one table called "Roads" with the following fields: ID, Road, Area, X and Y.

9.5 Object-oriented analysis and design

The city map applet with a road finder considered here should fulfill the following **requirements**. It should allow to:
1. View and scroll a city map,
2. Zoom in and out off a city map,
3. Search for a road in a city map and,
4. Move to and mark a road by a blinking text label like an "i" for information, if a search was successful.

A use-case diagram for these requirements is given in Fig. 9.3.

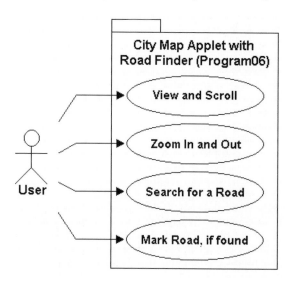

Fig. 9.3. Use-case diagram for a city map applet

As already mentioned in the last paragraph, the basic structure of the class diagram considered here is given through a generalization-specialization association between the *java.applet.Applet* and *Program06* classes. Further, other classes are added to *Program06* through a whole-to-part association, which can be categorized as follows:

- Classes for graphical components in applet Program06, which are java.awt.GraphClass; GraphClass ∈ {Button, Label, Panel, Scrollbar, TextField} and borland.jbcl.control.TransparentImage,
- Classes for database components in applet Program06, which are borland.sql.dataset.DBClass; DBClass ∈ {Database, QueryDataSet, QueryResolver} and borland.jbcl.dataset.ParameterRow, and
- Class BlinkingThread as an extension of java.lang.Thread to display a blinking text label.

All these added classes have a whole-to-part association with class *Program06* as shown in the class diagram below.

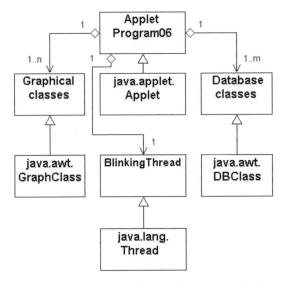

Fig. 9.4. Class diagram for the city map applet *Program06* (n = 6, m = 4)

The *java.awt* class package contains the most important classes for creating graphical components on a panel like on an applet's *java.awt.Panel*. Apart from the ones listed under *java.awt.GraphClass*, many more exist. For example, the *java.awt.Color* and *java.awt.Font* classes consist of attributes and operations for manipulating colors and fonts. Class *java.awt.Graphics* contains attributes and operations for drawing strings, lines, rectangles and other shapes. Class *java.awt.Toolkit* includes methods for getting graphical system information like the displayable fonts or the screen size.

The scenario for the use-case diagram in Fig. 9.3 is shown in the following sequence diagram (Fig. 9.5).

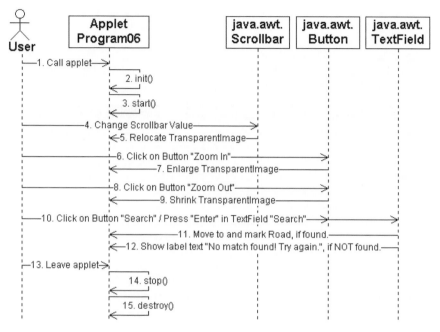

Fig. 9.5. Sequence diagram for the city map applet *Program06*

Here, a user calls an applet through an applet viewer or a Web browser in step 1. Each applet always begins with a series of three operation calls to "init()" in step 2, "start()" in step 3 and "paint()". Because the latter operation is not overridden in *Program06*, it was not included in this sequence diagram. In the end when the user leaves the applet in step 13, the "stop()"and "destroy()" operations are called to finish the applet's lifecycle in steps 14-15.

The "init()" operation is called once by the applet viewer or browser when an applet is loaded for execution. It initializes all applet objects. The "start()" operation is called after "init()" has completed its task in order to start the applet. This may include the display of graphical components or the start of an animation. The "stop()" operation is called when the applet should stop executing – normally when the Web page is left that contains the applet. Typical actions are then to stop animations and threads. The "destroy()" operation removes the applet from the computer memory – normally when a Web browser is closed.

9.6 Implementation of the designed model

The coded architecture of the city map applet with a road finder can be easily visualized with the JBuilder 2 project explorer (Jensen et al. 1998) as shown in the figure below. This window is automatically displayed after opening the application's project file named "Program06.jpr" in folder "Program06". All example programs incl. image and database files can be **downloaded** at http://de.geocities.com/bsttc2/book/SMOP.zip

Fig. 9.6. JBuilder project explorer window for the city map applet with source view

As can be seen in the top left corner of this figure, the project "Program06" in file named "Program06.jpr" consists of just two files "Program06.html" and "Program06.java". In this implementation, the names used in the sequence diagram

(Fig. 9.5) were retained as much as possible to make reading and understanding easier.

How project "Program06" works in detail and how it was coded, can be more easily described by going through the application's sequence diagram (Fig. 9.5) and explaining step-by-step the corresponding program code. Before class *Program06* can be initialized and started in steps 2-3 of Fig. 9.5, all required classes inside of *Program06* have to be declared. They are visible in the source view on the right-handed side of Fig. 9.6 and listed in the following table.

Table 9.4. Class declarations in file "Program06.java" in folder "Program06"

	...
01	XYLayout xYLayout1 = new XYLayout();
02	XYLayout xYLayout2 = new XYLayout();
03	Label label1 = new Label();
04	Label label2 = new Label();
05	Panel panel1 = new Panel();
06	Scrollbar scrollbarX = new Scrollbar();
07	Scrollbar scrollbarY = new Scrollbar();
08	TransparentImage transparentImage = new TransparentImage();
09	Button buttonZoomIn = new Button();
10	Button buttonZoomOut = new Button();
11	Button buttonSearch = new Button();
12	Button buttonBlink = new Button();
13	TextField textZoom = new TextField();
14	TextField textSearch = new TextField();
15	Label textNegativeResult = new Label();
16	BlinkingThread blink;
17	
18	boolean boolBlink;
19	int bx, by, sx, sy;
20	double zoom;
	...

Lines 01-02 declare two *XYLayout* named "xYLayout1, -2" objects for classes *Program06* and the *java.awt.Panel* object named "panel1" (line 05), which holds a *java.awt.Label* object named "label2" (line 04) and a *borland.jbcl.control.TransparentImage* object (line 08) as shown in Fig. 9.7.

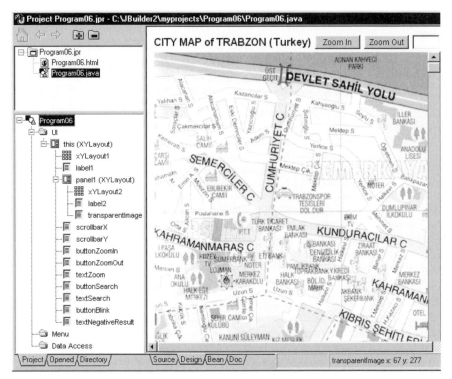

Fig. 9.7. JBuilder project explorer window for the city map applet with design view

The *java.awt.Label* object "label2" is used to display a blinking text label when a road was found after a successful search. The *borland.jbcl.control. TransparentImage* object named "transparentImage" (line 08 in Table 9.4) is placed on top of "panel1" so that "label2" becomes visible because of the transparency of "transparentImage". This is basically how step 11 in the sequence diagram (Fig. 9.5) is implemented.

Lines 03-15 declare other *java.awt.GraphClass* objects needed for the task at hand. Line 16 defines a *BlinkingThread* object named "blink", which will be described later in more detail. Lines 18-20 show definitions of other variables. "boolBlink" is a Boolean variable indicating whether "label2" is visible or not. "bx, by" are the present position values of the *java.awt.Scrollbar* objects named "scrollbarX" and "scrollbarY" (line 06-07 in Table 9.4). "sx, sy" are the X and Y coordinates of a road stored in the database. "zoom" is the present zoom factor of "transparentImage". Its default value is 100.

The initialization and start of *Program06* in steps 2-3 of the sequence diagram (Fig. 9.5) is done by the following code in Table 9.5, whereby the "init()" operation is replaced by a "jbInit()" operation as part of the *Program06* constructor in lines 01-08 in Table 9.5.

Table 9.5. Class initialization in file "Program06.java" in folder "Program06"

```
    ...
01  public Program06() {
02       try {
03         jbInit();
04       }
05       catch(Exception e) {
06         e.printStackTrace();
07       }
08  }
09
10  private void jbInit() throws Exception {
11      ...
12       boolBlink = false;
13       bx = scrollbarX.getValue();
14       by = scrollbarY.getValue();
15       zoom = 100;
16       textZoom.setText(DouToStr(zoom) +" %");
17       xYLayout1.setWidth(800);
18       xYLayout1.setHeight(600);
19       this.setLayout(xYLayout1);
20       this.setBackground(Color.white);
21
22       this.add(label1, new XYConstraints(1, 1, 336, 26));
23       this.add(panel1, new XYConstraints(2, 28, 540, 540));
24       this.add(scrollbarX, new XYConstraints(2, 568, 560, 20));
25       this.add(scrollbarY, new XYConstraints(542, 28, 20, 540));
26       this.add(buttonZoomIn, new XYConstraints(367, 3, 70, 21));
27       this.add(buttonZoomOut, new XYConstraints(443, 3, 70, 21));
28       this.add(textZoom, new XYConstraints(519, 3, 44, 22));
29       this.add(buttonSearch, new XYConstraints(570, 3, 70, 21));
30       this.add(textSearch, new XYConstraints(647, 3, 130, 22));
31       this.add(buttonBlink, new XYConstraints(-570, 29, 70, 21));
32       this.add(textNegativeResult, new XYConstraints(-647, 29, 151, 18));
33       panel1.add(label2, new XYConstraints(-100, -100, 10, 37));
34       panel1.add(transparentImage, new XYConstraints(0, 0, 600, 858));
35
36  } // END of private void jbInit() throws Exception {
    ...
```

In lines 12-20 in Table 9.5, variables "boolBlink", "bx, by", "zoom", "xYLayout1" and "this" are initialized. "this" represents the *Program06* object in the whole program. Because its layout is set to "xYLayout1" in line 19 and the size of "xYLayout1" is 800x600 pixel (lines 17-18), the size of *Program06* is also 800x600 pixel. Lines 22-34 add the different objects to the *Program06* object and "panel1" (lines 33-34), whereby their position and size are determined by the four arguments of the *XYConstraints* class: left position, top position, width and height (all in pixels).

A change of a scrollbar value causing a relocation of "transparentImage" on "panel1" according to steps 4-5 of the sequence diagram (Fig. 9.5) is coded as shown in the table below.

Table 9.6. Mouse moved operations of *java.awt.Scrollbar* objects in file "Program06.java" in folder "Program06"

```
        ...
01  void scrollbarX_mouseMoved(MouseEvent e) {
02      if (bx != scrollbarX.getValue()) {
03          System.out.println("Horizontal Value = " + scrollbarX.getValue() );
04          bx = scrollbarX.getValue();
05          transparentImage.setLocation(-bx,-by);
06          if (boolBlink == true)
07                  label2.setLocation(DouToInt(IntToDou(sx)*zoom/100) -bx,
08                                     DouToInt(IntToDou(sy)*zoom/100) -by);
09      };
10  }
11  void scrollbarY_mouseMoved(MouseEvent e) {
12      if (by != scrollbarY.getValue()) {
13          System.out.println("Vertical Value = " + scrollbarY.getValue() );
14          by = scrollbarY.getValue();
15          transparentImage.setLocation(-bx,-by);
16          if (boolBlink == true)
17                  label2.setLocation(DouToInt(IntToDou(sx)*zoom/100) -bx,
18                                     DouToInt(IntToDou(sy)*zoom/100) -by);
19      };
20  }
        ...
```

Lines 01-10 and 11-20 are identical for both java.awt.Scrollbar objects named "scrollbarX" and "scrollbarY" (line 06-07 in Table 9.4). If a scrollbar value has detected in line 02 or 12, its new value is printed in the console window (lines 03+13), stored in the variables "bx, by" (lines 04+14) followed by a relocation of "transparentImage" (lines 05+15). If the blinking "label2" is visible, it is also relocated accordingly in lines 06-08 and 16-18 using the X and Y coordinates of a road "sx, sy" stored in the database. Operations "DouToInt" and "IntToDou" are needed in lines 07-08 and 17-18 to transform double and integer formats of the same numbers.

Because zooming in or out off "transparentImage" in steps 6-7 and 8-9 of the sequence diagram (Fig. 9.5) is coded in an analog way, only the first operation will be described here using the following table.

Table 9.7. Button Zoom In operation in file "Program06.java" in folder "Program06"

```
     ...
01   void buttonZoomIn_actionPerformed(ActionEvent e) {
02       if (zoom < 500) {
03           zoom = zoom * 1.25;
04           textZoom.setText(DouToStr(zoom) +" %");
05           System.out.println(DouToStr(zoom) +" %");
06
07           Dimension d = transparentImage.getSize();
08           int w = DouToInt(d.width*1.25), h = DouToInt(d.height*1.25);
09           int x = scrollbarX.getMaximum(), y = scrollbarY.getMaximum();
10           System.out.println(scrollbarX.getValue() +" Old Value-X");
11
12           double xx = IntToDou(scrollbarX.getValue()+10)/
13                   IntToDou(x)*IntToDou(w - 540);
14           scrollbarX.setMaximum(w - 540);
15           scrollbarX.setValue(DouToInt(xx-10));
16           bx = scrollbarX.getValue();
17           System.out.println(scrollbarX.getValue() +" New Value-X");
18           System.out.println(scrollbarY.getValue() +" Old Value-Y");
19
20           double yy = IntToDou(scrollbarY.getValue()+10)/
21                   IntToDou(y)*IntToDou(h - 540);
22           scrollbarY.setMaximum(h - 540);
23           scrollbarY.setValue(DouToInt(yy-10));
24           by = scrollbarY.getValue();
25           System.out.println(scrollbarY.getValue() +" New Value-Y");
26
27           transparentImage.setBounds(-bx, -by, w, h);
28           if (boolBlink == true)
29               label2.setLocation(DouToInt(IntToDou(sx)*zoom/100) -bx,
30                           DouToInt(IntToDou(sy)*zoom/100) -by);
31       }
32   }
     ...
```

After checking the "zoom" factor in line 02 in Table 9.7, which should be smaller than 500, the new "zoom" factor is computed in line 03 and displayed in lines 04-05. In line 07, the present size of "transparentImage" is stored in a *Dimension*

mension object named "d", keeping its width and height. The enlarged width and height are computed with a multiplication factor of 1.25 and stored in "w, h" in line 08.

The problem is now to update the scrollbar positions and maximum values accordingly. For this, the present scrollbar positions are stored in "x, y" in line 09. The new maximum values "xx, yy" are computed in lines 12-13 and 20-21, whereby an offset of 10 pixels need to be taken into account, and the actual size of "panel1" of 540 pixels must be subtracted (line 23 in Table 9.5). Then the scrollbar positions and maximum values are set to its new values in lines 14-16 and 22-24. Afterwards, "transparentImage" is relocated in lines 27. If the blinking "label2" is visible, it is also relocated accordingly in lines 28-30.

The search operation in steps 10-12 of the sequence diagram (Fig. 9.5) is coded as shown in the following table.

Table 9.8. Button Search operation in file "Program06.java" in folder "Program06"

```
...
01  void buttonSearch_actionPerformed(ActionEvent e) {
02      textSearch.requestFocus();
03      textNegativeResult.setLocation(-647, 29);
04
05      String s = SearchRoad(textSearch.getText());
06      if (s.equals(textSearch.getText())== true) {
07
08          buttonBlink.setLocation(570, 29);
09          Dimension d = transparentImage.getSize();
10          int w = DouToInt(d.width), h = DouToInt(d.height);
11          int x = scrollbarX.getMaximum(), y = scrollbarY.getMaximum();
12          System.out.println(sx +" Search Value-X");
13          System.out.println(scrollbarX.getValue() +" Old Value-X");
14
15          double xx = IntToDou(sx)/IntToDou(w)*IntToDou(w - 540)*zoom/100;
16          scrollbarX.setValue(DouToInt(xx));
17          bx = scrollbarX.getValue();
18          System.out.println(scrollbarX.getValue() +" New Value-X");
19          System.out.println(sy +" Search Value-Y");
20          System.out.println(scrollbarY.getValue() +" Old Value-Y");
21
22          double yy = IntToDou(sy)/IntToDou(h)*IntToDou(h - 540)*zoom/100;
23          scrollbarY.setValue(DouToInt(yy));
24          by = scrollbarY.getValue();
25          System.out.println(scrollbarY.getValue() +" New Value-Y");
26
```

```
27        transparentImage.setBounds(-bx, -by, w, h);
28        if (boolBlink == true)
29            label2.setLocation(DouToInt(IntToDou(sx)*zoom/100) -bx,
30                              DouToInt(IntToDou(sy)*zoom/100) -by);
31        boolBlink = true;
32        blink = new BlinkingThread(700);
33        blink.start();
34
35    } else { System.out.println("No match found! Try again.");
36            textNegativeResult.setLocation(647, 29); this.repaint();
37            buttonBlink_actionPerformed(e);
38    }
39 }
          ...
```

The code in Table 9.8 is very similar to the one in Table 9.7. The main difference is the "If-Then-Else" clause starting in line 06 and ending in line 38, which checks whether a road name given by *java.awt.TextField* object "textSearch" is existing in the database. If yes, the returned String "s" from operation "SearchRoad" in line 05 is identical to the text in "textSearch" and lines 08-33 are executed. If not, the *java.awt.Label* object "textNegativeResult" is shown (line 36) and the blinking text label "label2" is closed, if visible (line 37). What is left here to describe is the new code in Table 9.8: the operation "SearchRoad" in line 05 and the *BlinkingThread* object named "blink" in line 32. The code for "SearchRoad" is listed in the table below.

Table 9.9. "SearchRoad" operation in file "Program06.java" in folder "Program06"

```
      ...
01 public String SearchRoad(String searchText) {
02        Database db = new Database();
03        QueryDataSet qds = new QueryDataSet();
04        QueryResolver qr = new QueryResolver();
05        ParameterRow pr = new ParameterRow();
06        String s = "", TableName = "Roads";
07
08 try {
09        db.setConnection(new borland.sql.dataset.ConnectionDescriptor(
10          "jdbc:odbc:CityMap", "", "", false, "sun.jdbc.odbc.JdbcOdbcDriver"));
11
12        Column c0 = new Column(); c0.setColumnName("ID");
13        c0.setDataType(borland.jbcl.util.Variant.INT);
14        qds.setColumns(new Column[] {c0} );
15        qds.setRowId("ID", true);
16        qds.setTableName("["+TableName+"]");
```

```
17          qds.setResolver(qr);
18
19  pr.addColumn(new Column("Text1","", borland.jbcl.util.Variant.STRING) );
20  pr.addColumn(new Column("Text2","", borland.jbcl.util.Variant.STRING) );
21          pr.setString("Text1", searchText + " C");
22          pr.setString("Text2", searchText + " S");
23
24          qds.setQuery(new borland.sql.dataset.QueryDescriptor(db,
25            "SELECT * FROM "+ "["+TableName+"] WHERE "+
26                          "(["+TableName+"].Road = :Text1)"
27              +" OR "+ "(["+TableName+"].Road = :Text2)",
28          pr, true, Load.ALL));
29          qds.executeQuery();
30          qds.open();
31
32          if ( qds.getRowCount() > 0 ) { s = searchText;
33              sx = new Integer(qds.getString("X")).intValue()-15;
34              sy = new Integer(qds.getString("Y")).intValue()-20; }
35      }
36        catch (Exception ex) {
37          DataSetException.handleException(ex);
38        }
39    return s;
40  } // END of public String SearchString(String searchText)
        ...
```

The "SearchRoad" operation in Table 9.9 is coded as an independent module where everything including the class declarations and initializations are located inside of this operation. Only the two required packages

```
06 | import borland.jbcl.dataset.*
07 | import borland.sql.dataset.*
```

have to be imported at the beginning of *Program06*. The *String* object "searchText" is passed to the operation in line 01 and the resulting *String* object "s" is passed back to the calling object in line 39. The search operation needs four data access objects defined in lines 02-05:

- a borland.sql.dataset.Database object named "db",
- a borland.sql.dataset.QueryDatSet object named "qds",
- a borland.sql.dataset.QueryResolver object named "qr", and
- a borland.jbcl.dataset.ParamaterRow object named "pr".

The *Database* object "db" provides the connection to a database and has to be included. It uses a "setConnection()" operation (line 09 in Table 9.9), which takes as argument a *borland.sql.dataset.ConnectionDescriptor* object. This object has the following parameters (line 10): connection name, user name, password, prompt for password as Boolean, and database driver (Jensen et al. 1998).

The *QueryDataSet* object "qds" allows querying a database using SQL statements (see paragraph 8.1). Before this can happen, a *borland.jbcl.dataset.Column* object named "c0" is created with a label "ID" and set as column to "qds" to be able to query the table provided by "TableName" (lines 12-16).

The *QueryResolver* object "qr" provides control over the resolve process when querying a database and is set to "qds" in line 17. For example, "qr" includes an "UpdateMode" attribute to store data changes. This attribute can be set to all columns, changed columns, and key columns or left unassigned.

The *ParameterRow* object "pr" is used in what is called a "parameterized query". This is a query with variables or parameters like in lines 26-27, where two string columns named "Text1" and "Text2" defined in lines 19-20 and initialized in lines 21-22, are applied to search for a road in "TableName" in a column named "Road". Hereby, the "searchText" variable just requires the name. Its extensions " C" or " S" indicate a larger ("Caddesi") or smaller road ("Sokak") and are automatically added in lines 21-22.

Then the *QueryDataSet* object "qds" is executed and opened in lines 29-30. If the road name "searchText" exists, "qds" is not empty and the "getRowCount()" operation is larger than null (line 32). In this case, string "s" is set to "searchText" and the X and Y coordinates "sx, sy" of a road stored in the database are assigned with offsets of minus 15 and 20 pixels respectively (lines 32-34).

The *BlinkingThread* object named "blink" in line 32 in Table 9.8 is given by the code in the following table.

Table 9.10. Class *BlinkingThread* in file "Program06.java" in folder "Program06"

```
      ...
01  class BlinkingThread extends Thread {
02          long sleepTime;
03          BlinkingThread(long sleepTime) {
04            this.sleepTime = sleepTime; }
05    public void run() {   // sleeps for N millisecondes
06
07        try{ int i = 1;
08        while ( i < 10 ) {
09        if (i==1 || i==3 || i==5 || i==7) label2.setText("");
10        else label2.setText("i");
11        this.sleep(sleepTime);
12        i++; if (i == 9) i = 1;  } // END of while()
13        } catch (Exception ex) {
```

```
14 |        System.out.println(ex);
15 |      } // END of try
16 |    } // END of run()
17 | } // END of class BlinkingThread extends Thread {
   |    ...
```

Class *BlinkingThread* is derived from class *java.lang.Thread* by a generalization-specialization association in line 01. After starting the *BlinkingThread* object "blink" in line 33 in Table 9.8, the "run()" operation of "blink" is executed in lines 05-16. It contains an unending while loop in lines 08-12 that waits for the given sleeping time (here 700 milliseconds or 0.7 seconds) at each step before "blink" is stopped through its "stop()" operation. Depending on the integer value "k", the *java.awt.Label* object "label2" shows an "i" or an empty text string (lines 09-10), thus causing a blinking text label.

Multi-threading based on class *java.lang.Thread* is the main topic of the next application in chapter 10. Then Java threads will be discussed in more detail including their advantages and disadvantages.

9.7 Patterns used in this application

As already mentioned in paragraph 9.4, the 3-tier architecture used so far was abandoned here and replaced by a single module and Java file called "Program06.java". Nevertheless, the following **design patterns** are used because classes and objects in Java behave by their nature according to these patterns.

Creational Patterns:

2. Builder: Separate the construction of a complex object from its representation so that the same construction process can create different representations.

5. Singleton: Ensure a class only has one instance, and provide a global point of access to it.

Structural Patterns:

6. Adapter: Convert the interface of a class into another interface clients expect. Adapter lets classes work together that couldn't otherwise because of incompatible interfaces.

Behavioral Patterns:

14. Command: Encapsulate a request as an object, thereby letting you parameterize clients with different requests, queue or log requests, and support undoable operations.

19. Observer: Define a one-to-many dependency between objects so that when an object changes state, all its dependents are notified and updated automatically.

The **builder and command patterns** are used in this application when zooming in or out off *borland.jbcl.control.TransparentImage* object "transparentImage" (Table 9.7). Instead of two different operations for a key press and a mouse click, which provide exactly the same functionality, only one reusable object of class *ZoomAction* can be applied to have just one single command, just simplifying the code. For this, the constructor of class *ZoomAction* creates different objects depending on just one parameter, a multiplication factor for the "zoom" factor. To implement class *ZoomAction*, it must be added to the class declarations of class *Program06* in Table 9.4 as follows:

17 | ZoomAction zAction;

Then the code for all zoom in and –out operations can be simplified as shown in the table below.

Table 9.11. Zoom In and –Out operations in file "Program06.java" in folder "Program06"

```
       ...
01     void buttonZoomIn_actionPerformed(ActionEvent e) {
02         zAction = new ZoomAction(1.25);
03     }
04     void buttonZoomIn_keyPressed(KeyEvent e) {
05      if (e.getKeyText(e.getKeyCode()).equals("Enter")==true)
06         zAction = new ZoomAction(1.25);
07     }
08     void buttonZoomOut_actionPerformed(ActionEvent e) {
09         zAction = new ZoomAction(0.8);
10     }
11     void buttonZoomOut_keyPressed(KeyEvent e) {
12      if (e.getKeyText(e.getKeyCode()).equals("Enter")==true)
13         zAction = new ZoomAction(0.8);
14     }
15 class ZoomAction { ZoomAction(double fac) {
16    if (fac > 1 && zoom < 500 || fac < 1 && zoom > 100) {
17       zoom = zoom * fac;
18       textZoom.setText(DouToStr(zoom) +" %");
```

```
19              System.out.println(DouToStr(zoom) +" %");
20
21              Dimension d = transparentImage.getSize();
22              int w = DouToInt(d.width*fac), h = DouToInt(d.height*fac);
23              int x = scrollbarX.getMaximum(), y = scrollbarY.getMaximum();
24              System.out.println(scrollbarX.getValue() +" Old Value-X");
25
26              double xx = IntToDou(scrollbarX.getValue()+10)/
27                      IntToDou(x)*IntToDou(w - 540);
28              scrollbarX.setMaximum(w - 540);
29              scrollbarX.setValue(DouToInt(xx-10));
30              bx = scrollbarX.getValue();
31              System.out.println(scrollbarX.getValue() +" New Value-X");
32              System.out.println(scrollbarY.getValue() +" Old Value-Y");
33
34              double yy = IntToDou(scrollbarY.getValue()+10)/
35                      IntToDou(y)*IntToDou(h - 540);
36              scrollbarY.setMaximum(h - 540);
37              scrollbarY.setValue(DouToInt(yy-10));
38              by = scrollbarY.getValue();
39              System.out.println(scrollbarY.getValue() +" New Value-Y");
40
41              transparentImage.setBounds(-bx, -by, w, h);
42              if (boolBlink == true)
43                      label2.setLocation(DouToInt(IntToDou(sx)*zoom/100) -bx,
44                              DouToInt(IntToDou(sy)*zoom/100) -by);
45      }
46 } // END of class ZoomAction
   ...
```

Lines 05 and 12 check whether the pressed key is the "Enter" key using the "getKeyText(e.getKeyCode())" operation of a *KeyEvent* object "e". After checking the multiplication factor "fac" and the "zoom" factor in line 16, which should be larger than 100 and smaller than 500, the identical operation is computed as in Table 9.7.

The Singleton creational pattern is used in every applet because of class *java.applet.Applet*. This pattern ensures that only one object of the applet class, here *Program06*, exists in an applet.

The Adapter structural pattern and the Observer behavioral pattern are used, for example, when *java.awt.Scrollbar* objects respond to *MouseEvent* objects using *MouseMotionListener* and *MouseMotionAdapter* objects (see code line 01 below). The *MouseMotionListener* object observes *MouseEvent* objects through the *MouseMotionAdapter* object and informs the *Scrollbar* object permanently of any state changes of *MouseEvent* objects. The *MouseMotionAdapter* object makes sure that *Scrollbar* objects can communicate with *MouseEvent* objects.

```
01   scrollbarX.addMouseMotionListener(
02       new java.awt.event.MouseMotionAdapter() {
03           public void mouseMoved(MouseEvent e) {
04               scrollbarX_mouseMoved(e); }        });
```

9.8 Testing the application

The city map applet with a road finder required a lot of initial testing to try out different GUI components and decide then, which ones serve the given requirements in paragraph 9.5 the best. Moreover, the final archive of the *Program06* applet should be as small as possible to guarantee short applet loading times. Therefore, the basic GUI components like buttons and text fields were all taken from the *java.awt.** class library.

Instead of *java.awt.Panel* and *java.awt.Scrollbar* objects, for example, the *java.awt.ScrollPane* was used in the beginning because it was thought that one single component is better than two. But after a lot of trial-and-error it was discovered that basic operations of *java.awt.ScrollPane* do not work, as they should do. Then it was decided to go for *java.awt.Panel* and *java.awt.Scrollbar* objects, which delivered better results.

But also these components caused a lot of difficulties, but they could be solved at least. For example, *java.awt.Scrollbar* objects offer a lot of different operations than the "mouseMoved" operation listed in Table 9.6, such as "mouseClicked", "mousePressed" or "mouseReleased". But none of them worked satisfactorily. Only the "mouseMoved" operation gave good enough results. This is also the reason behind the code in Table 9.6 with the present position values of the *java.awt.Scrollbar* objects "bx, by", which allow to store these position values.

Table 9.12. Faulty class initialization in file "Program06.java" in folder "Program06"

	...
33	panel1.add(label2, new XYConstraints(-100, -100, 10, 37));
34	panel1.add(transparentImage, new XYConstraints(0, 0, 600, 858));
35	
36	} // END of private void jbInit() throws Exception {
	...

Another problem encountered was the sequence in which objects are added to the *Program06* object named "this" in Table 9.4. By swapping lines 33 and 34 as shown in the Table 9.12, for example, the blinking text label in "label2" may be covered by "transparentImage", thus becoming invisible, if the "transparent" attribute of "transparentImage" is set to false.

Line 33 in Table 9.12 points to another weak point of *java.awt.GraphClass* objects when initialized. The *java.awt.Label* object "label2" is added to "panel1" but with negative values for Left and Top coordinates (each - 100 pixel) so that it is not visible on "panel1". This was done because the following code:

Table 9.13. Faulty class initialization in file "Program06.java" in folder "Program06"

	...
33	panel1.add(label2, new XYConstraints(100, 100, 10, 37));
34	label2.setVisible(false);
35	panel1.repaint();
36	} // END of private void jbInit() throws Exception {
	...

and then setting at line 08 in Table 9.8 instead:

Table 9.14. Button Search operation in file "Program06.java" in folder "Program06"

	...
01	void buttonSearch_actionPerformed(ActionEvent e) {
02	textSearch.requestFocus();
03	textNegativeResult.setLocation(-647, 29);
04	
05	String s = SearchRoad(textSearch.getText());
06	if (s.equals(textSearch.getText())== true) {
07	
08	label2.setVisible(true);
09	panel1.repaint();
	...

does not result in a blinking text "label2". Therefore, this application was coded with negative values for Left and/or Top coordinates to make *java.awt.GraphClass* objects invisible. All applets were tested with MS Internet Explorer version 5.0.

The "setConnection()" operation in line 09 in Table 9.9 should be used with a Java Database Connectivity (JDBC) driver of Type 4 because they require no other installations and are pure Java solutions (Deitel et al. 2002). For example, JDBC database drivers are available for download at http://java.sun.com/ products/jdbc/.

JDBC drivers are divided into four types or levels. Each type defines a JDBC driver implementation with increasingly higher levels of platform independence, performance, and deployment administration. The four types are:
1. Type: JDBC-ODBC Bridge,
2. Type: Native-API/partly Java driver,
3. Type: Net-protocol/all-Java driver, and
4. Type: Native-protocol/all Java driver.

The type 1 driver, JDBC-ODBC Bridge, translates all JDBC calls into ODBC (Open Database Connectivity) calls and sends them to the ODBC driver. As such, the ODBC driver, as well as, in many cases, the client database code must be present on the client machine. In the present version of this application, a bridge pointing to the database file "CityMap.mdb" in folder "Program06" with the name "CityMap" (first parameter in line 10 in Table 9.9) has to be installed (Jensen et al. 1998).

The JDBC driver type 2 - the native-API/partly Java driver - converts JDBC calls into database-specific calls for databases such as SQL Server, Informix, Oracle, or Sybase. The type 2 drivers communicate directly with the database server; therefore it requires that some binary code be present on the client machine.

The JDBC driver type 3 - the net-protocol/all-Java driver - follows a three-tiered approach whereby the JDBC database requests are passed through the network to a middle-tier server. The middle-tier server then translates the request (directly or indirectly) to the database-specific native-connectivity interface to further the request to the database server. If the middle-tier server is written in Java, it can use a type 1 or type 2 JDBC drivers to do this.

The native-protocol/all-Java driver (JDBC driver type 4) converts JDBC calls into the vendor-specific database management system (DBMS) protocol so that client applications can communicate directly with the database server. Level 4 drivers are completely implemented in Java to achieve platform independence and eliminate deployment administration issues.

CHAPTER 10: Parallel Computing of Satellite Orbits using Java Threads

10.1 Introduction to parallel computing

10.1.1 Some fundamentals

Two qualifiers classify parallel computers (Ragsdale 1992): the relationship of the memory to the processors and the number of instruction streams available to the system. There are two general relationships between memory and processors: **distributed memory** and **shared memory**. In distributed-memory systems, each processor has its own private memory. Shared-memory systems have a single pool of memory to which all processors have access.

Other expressions used to describe distinguishing characteristics of parallel computers are **MIMD** (Multiple Instruction, Multiple Data) and **SIMD** (Single Instruction, Multiple Data). These acronyms are commonly pronounced "mim-dee" and "sim-dee". MIMD systems allow processors to work on separate instruction streams, or tasks, at the same time, while the processors in SIMD systems all operate on a single instruction stream simultaneously. Synchronous behavior is automatic in SIMD machines; MIMD systems require that synchronization be programmed in. A MIMD machine can simulate a SIMD machine, but not vice-versa. MIMD machines can make use of either shared or distributed memory, but only distributed memory is used in SIMD machines because multiple processors could not be used efficiently with a single instruction stream and a single memory pool.

MIMD systems consist of a set of nodes, each of which holds a main processor, memory, and interface to the network. Each node has the computing power of a stand-alone computer. Nodes process information independent of one another and communicate by sending and receiving messages. This independence gives these systems what is called a loosely coupled architecture. MIMD systems have three major **advantages**:

- They are flexible. Nodes with different types of processors can be used to adapt the system to specialized problems.
- They are cost-effective. Parallel processing can significantly reduce the processing time for large computation-intensive problems. While some systems use specially designed components, others further reduce costs by using existing technology wherever possible.
- They are easily scalable. Loosely coupled systems are easy to expand, and you can also run several smaller problems concurrently.

A program designer has to figure out the best way to cut (or decompose) the problem into pieces that can then be distributed among the nodes. It also enhances the variety of ways the computer can be used to solve problems, for example:
- A single program can be divided into different parts, with nodes executing different parts of the program (option A)
- Several separate programs can be executed at once on different sets of nodes (option B), or
- A single program can be executed simultaneously by different nodes emulating a single instruction, multiple data (SIMD) machine (option C).

This chapter includes two Java thread programs for:
- Option A: computing number π, and
- Option C: computing satellite orbits.

10.1.2 Communication and messages

The processors in a loosely coupled parallel system operate independently of one another. As a result, processes on different nodes are executed asynchronously with no guaranteed timing relationship between them. Communication serves both to synchronize processes and to exchange code and data information among processes (Ragsdale 1992).

Communication among processes is based on a very simple model. When a process requires information from another process, the process originating the information must send a copy of the information to the receiving process in the form of a message, and the receiving process must explicitly receive it. Documents are often sent before the receiver is ready. Perhaps more frequently, a document is required before it has arrived. Similar situations occur when passing messages between processes in parallel applications.

Synchronous message handling means the halting of further execution of instructions in a process until a message is sent or received. When a process issues a request to receive a message, the process executes no further instructions (is blocked) until a message of the type expected has been received. If the message has not yet arrived at the node at the time the request is made, the process will do nothing until the message arrives. Similarly, when a process makes a request to send a message synchronously, the process is blocked until the message is copied by the operating system from the sending process' memory into the message-passing network. This does not mean that the message has arrived at its destination. It means that it is on its way, and that the memory that stored the outgoing message can now be changed if needed.

Asynchronous message handling means, instead of waiting for an expected message, a process continues working, until the expected message arrives and the process is ready to receive it. It is usually more efficient to handle messages asynchronously. This type of message handling allows processes to alert the operating system that certain messages are expected and should be delivered to the process as soon as they arrive, even though the process may be busy doing other work at the time. The process does not idly wait for the message to arrive; it continues to execute instructions until the information in the message is required. Then, the process must check, if the message has been delivered. If it has, the process can use it immediately. If the message has not, the process, will, at that point, wait for it to arrive.

Interrupt message handling is similar to asynchronous message handling but, while working and waiting for a message to arrive, the process receives an interrupt as soon as the message arrives so it can be processed immediately. Afterwards, the process can return to what it was doing when the interrupt occurred.

10.1.3 Performance

Some general terms are used to describe the performance of multiprocessor systems. The two most commonly used terms are speed-up and efficiency. **Speed-up** describes parallel performance as follows:

$$S_P = T_1 / T_P,$$

where S_P is the speed-up on p nodes, T_1 the time required using the best sequential algorithm to solve the problem, and T_P the time required for the parallel algorithm to solve the equation on p nodes. Linear speed-up would mean that the time required to run the problem on one node would be p times the time required to run it on p nodes. **Efficiency** is defined as speed-up per node:

$$E_P = S_P / p.$$

The performance advantage you get by using a parallel computer over a sequential computer depends upon the type of problem you are solving and the way you have broken the application up into processes (decomposed it). Overhead introduced by the parallel decomposition can adversely affect performance if not done carefully. For example, you would expect linear speed-up from the kind of problem where the processes running on different nodes are simply different cases of a single program, so concerns about breaking up sequential portions and communication issues do not arise. On the other hand, if you are running a complex single program as a set of processes, it might be more difficult to obtain strictly linear speed-up, although there is no question that the performance would be significantly enhanced.

10.1.4 Decomposition strategies

With a programming model in mind, the next step is to understand the process of producing a parallel program. This includes three major steps (Ragsdale 1992):
1. Decomposition - the division of the application into a set of parallel processes and data.
2. Mapping - the way processes and data are distributed among the nodes.
3. Tuning - alteration of the working application to improve performance.

Inherent in this process are certain principles that are guides to designing an efficient program. These principles help you to determine each of the steps of the process:
- Balance the computational load,
- Minimize the communication to computation ratio,
- Reduce sequential bottlenecks, and
- Make the program scalable (independent of the actual number of available nodes).

This paragraph introduces the decomposition process followed by an example: a partial Gaussian elimination process with objects. In general, there are three decomposition methods existing:
1. Perfectly parallel decomposition,
2. Domain decomposition, and
3. Control decomposition.

1. Perfectly parallel decomposition is the easiest technique and can be applied to problems that can be divided into a set of processes that require little or no communication with one another. The calculation of π is a simple example of a perfectly parallel application. The calculation of the areas of rectangles is divided up among the available processors, and only the results need to be communicated. Further, the order in which the areas of the rectangles are calculated is not important, so no synchronization is required during the calculations.

2. Domain decomposition is generally applied to systems with a large, discrete, static data structure. Decomposing the "domain" of computation, the fundamental data structures, provides the road map for writing the program. Domain decomposition works particularly well for the following three kinds of problems:

- Problems where the data structure is static, for example, factoring and solving a large matrix or finite difference calculations on a mesh are both natural candidates for domain decomposition.
- Problems where the data structure is dynamic, but it is tied somehow to a single entity. For example, in a large multi-body problem, some subset of the bodies could be distributed to each node. Even though the bodies might be moving through space and interacting with each other, the calculations for each body can stay on the original node.
- Problems where the domain is fixed but the computation within various regions of the domain is dynamic. An example of this kind of application is a program that models fluid vortices, where the domain (say, a section of the Gulf Stream) stays fixed but the whirlpools move around.

3. Control decomposition can be used for problems where the main characteristic is the flow of control, such as a power plant or an airplane's flight control system. One possible control decomposition technique is called manager/worker. This technique uses one process as the "manager", which farms out the tasks (processes) to each of the "workers" (nodes) and keeps track of the progress of the computation as the workers report back with completed tasks.

All decomposed systems and parallel programs consist of multiple threads of execution that access both private and shared data. Data values are very naturally represented as attributes of object. Threads of execution can be allocated as task type objects. If task objects must communicate by exchanging messages, they include operations for sending and receiving messages. Data objects that are private to a task interact with one another via direct invocation of a private object's functional interface. Shared data structures can be organized as objects whereby multiple task objects can be synchronized by queue objects. Object orientation offers certain advantages for parallel systems. It inherently avoids the use of global variables, simplifying the job of partitioning data and code across the processing nodes. In addition, the use of operations to access objects provides a higher granularity of interaction than direct access to object variables. This can improve the efficiency of the underlying message-passing system.

10.1.5 An example: partial Gaussian elimination

Partial gaussian elimination is a technique for factoring a matrix in linear equations. A sequential algorithm can be summarized as follows (Ragsdale 1992):

```
for (all columns i)
    Select_max (column i);
    for (all columns j > i)
        Transform (column j using column i)
    end for
end for
```

The "Select_max" operation encapsulates several operations, including selecting the maximum value in column i, swapping this element with the i-th element, and normalizing all elements. The "Transform" operation swaps the same two element positions in the j-th column (thus swapping two rows of the matrix in total), and then it transforms the values of column j using values from column i.

This algorithm manipulates the columns of a matrix. A matrix is represented as a one-dimensional array of columns. Therefore, it is convenient to define the object type column associated with the operations "Select_max" and "Transform" to manipulate objects of this type. You would also need other operations to initialize and print the values within a column. By defining the "Select_max" and "Transform" operations on column objects, you can define precisely how various portions of the algorithm manipulate the column data.

A parallel version of the sequential algorithm above is a straight-forward extension of the same algorithm and reads as follows:

```
for (all columns i)
    Select_max (column i);
    Do in parallel for (all columns j > i)
        Transform (column j using column i)
    end for
end for
```

Here the difference is that the "Transform" procedure is executed in parallel for all columns j > i. The first task is to partition the columns of the matrix across the processing nodes. The "Transform" operation is executed on all of the column objects in parallel. To eliminate the need for global synchronization between steps of the inner for-loop, wrapping the "Select" and "Transform" procedures within two new procedures operating on a column object, the algorithm can be reorganized as follows:

```
Select_and_transform (column i) {
    Select_max (column i);
    Do in parallel for (all columns j > i)
        Transform_and_select (column j using column i, pivot)
    end do
}

Transform_and_select (column j using column i, pivot) {
    Transform (column j using column i, pivot);
    if (i < N) {
        Select_and_transform (column i+1) {
    }
}
```

Execution of this algorithm is initiated by invoking the operation "Select_and_transform" on column 1 of N columns. Program execution is carried out as a sequence of operations on column objects, most of which are executed in parallel when using the "Transform_and_select" operation inside of the do in parallel for-loop.

The difference to a non-object-oriented algorithm is the explicit definition of column objects and its focus on the "logical" parallelism provided in transforming these objects. The object-oriented style explicitly describes the data structures and the operations that operate on them. It focuses on objects as logical entities and the behavior and interaction of those objects.

10.2 Introduction to multi-threading in Java

10.2.1 Some fundamentals

A thread is **defined as** a sequence of code statements executed when a thread is started. Programming languages like Java support multiple threads that can run concurrently, meaning that a program can be doing a number of different things at once. But threads can only be executed in parallel, if the program is running on a multi-processor machine so that each processor executes a thread at the same time. A single-processor machine only appears to be running multiple threads in parallel because it can only execute one instruction from one thread at any given time. Multiple threads are interleaved on this type of machines, meaning that the currently executing thread is periodically switched for another one (Winder & Roberts 1998).

There are many different **reasons why** a program is in need of multiple threads. In general, multi-threading helps to improve program performance by eliminating time delays and using idle time. For example:
- When saving user input to a database, an application can continue to accept other data input.
- When downloading and processing large files, an application can already start to process the file during the download after a sufficient amount of data was received.
- When processing lengthy requests, a server application can continue to service other requests before the original transaction is completed.
- When printing documents, an application can allow further browsing and editing the document.
- When using a timer like for the blinking text label in Table 9.10, an application can start a timer thread (class *BlinkingThread* in Table 9.10) and at the same time is free to do other work, such as scrolling and zooming an image.

Another example of multi-threading is Java's automatic garbage collection. Java includes a garbage collector thread that automatically and continuously reclaims dynamically allocated memory that is no longer needed.

Multi-threading is also commonly used for managing user interfaces containing windows, menus, buttons, list or text boxes, etc. Any program should have a user interface that responds quickly to user input like a key press or a mouse click. For such requirements a program is hard to create without multi-threading, especially if the program is busy with ongoing processing. Then it might ignore the user interface, giving the impression that it had crashed. Or the program needs to include loops that check the user interface for input. A multi-threaded program is able to better solve these problems by using threads to manage the user interface while other threads are responsible for other tasks.

On the other hand, multi-threading does not allow individual tasks to be completed any faster on single-processor machines. Just the opposite, such tasks like rendering a graphic or a numerical computation is completed in the same time or even slower on one processor because the thread creation and management (mainly their scheduling and synchronization) take also a certain time. As mentioned before, the main **benefits** of multi-threading originate from the overlapping of different tasks so that time delays are shortened and idle time not wasted. This is done by an adequate decomposition of the problem at hand and distributing these components to different threads (see the examples listed before).

Besides the slower run-time for multi-threaded programs on single-processor machines, there are other **disadvantages** of multi-threading existing. One of them is a practical limit of how many threads a computer and its operating system can effectively manage. When dealing with a larger number of threads, the performance of many machines drops significantly. In practice, smaller applications should only use a few threads, and even large programs should not run more than 20 threads at the same time (Winder & Roberts 1998).

The other disadvantage of multi-threading is a more theoretical one and related to the synchronization of threads. Threads that are created within one program can gain access to any object or its data in this program. This can lead to unpredictable problems when not handled with care. For example, what happens when two threads try to update the same data at the same time? This can lead to unforeseeable results as long as there are no rules implemented to monitor, lock and synchronize threads in such situations. Thus, the development of thread-safe applications requires more effort and time.

Further, the inclusion of monitors and locking mechanisms to prevent threads from manipulating the same objects or their data at the same time has negative effects on the performance of a program. And there is the potential that another famous problem may occur: a deadlock, meaning that one or more threads continue to run or wait until an event takes place that will never happen in practice.

The Java solution to these kinds of problems are the so called "monitors", which are high-level mechanisms for ensuring that only one thread at a time executes a critical region of code. The code comprising the critical region is guaranteed by the monitor to be executed by one and only one thread at any time. If a thread tries to enter a critical region that is already in use by another thread, the monitor will block it until the other thread leaves the region. The implementation of monitors and hence the enforcement of access to critical regions is provided by locks on objects, so each critical region is associated with an object that has a lock. When a thread enters the critical region it tries to obtain the object lock and if successful can proceed into the critical region. The lock is released when the thread leaves the critical region (Winder & Roberts 1998).

10.2.2 Thread management

Monitors are one very important management tool of Java threads. Other tools were already introduced in chapter 9 of this book when describing the functionality of thread operations like "start", "run" or "stop" (Table 9.10). There are many more attributes and operations of class *java.lang.Thread* existing for creating and managing threads, which will be explained in the next paragraph about thread syntax. Here it is enough to stress that an application must have all the tools to manage its threads effectively. For example, it needs to be able to control which thread is presently running and how to switch from one thread to another. If such thread control is not available to a program, the benefits of using threads are lost. An important part of thread control is the priority level given to a thread. A thread with a higher priority level will be executed before a thread with a lower priority. In summary, multi-threading is a very convenient and versatile method for helping to create more high-quality applications under the condition that proper management and synchronization rules are applied.

10.2.3 Thread syntax

Class *java.lang.Thread* provides a variety of attributes and operations to create, manage and destroy Java threads. Because this class is a part of the core Java library "java.lang", it is guaranteed that Java threads can run and be used on any computer that has a Java Virtual Machine (JVM) installed. In the Java programming language, threads are objects of class *java.lang.Thread*. If a new thread is needed, a new instance of class *java.lang.Thread* is created. What follows is a description of important features of class *java.lang.Thread*. A **full documentation** can be found at http://java.sun.com/products/jdk/1.1/docs/api/java.lang.Thread. html. The same HTML document is also included in folder "Program07" of the example programs.

When created all *java.lang.Thread* objects or shorter *Thread* objects have a **name**. If not set explicitly in the program, a default name is automatically generated by the JVM. By default, each *Thread* object is a member of a *ThreadGroup* object, which can also be named explicitly, if the default name of the *ThreadGroup* object is not convenient. There are two common ways to start Java *Thread* objects:

1. Create a new thread by a generalization-specialization association with class *java.lang.Thread* and include a self-written "run()" operation (lines 03-05 in the code example below). When the MyThread object named "thread1" is created (line 09) and started by calling the "start()" operation (line 10), it begins executing by first calling the "run()" operation in lines 03-05. This was exactly done in Table 9.10 and is illustrated here again with the following code example:

```
01 | class MyThread extends Thread
02 | { ...
03 | public void run()
04 | { ... // Include self-written code here
05 | }
06 | ...
07 | }
08 | ...
09 | MyThread thread1 = new MyThread();
10 | thread1.start();
11 | ...
```

2. Create a new thread by implementing the Java "Runnable" interface and include a self-written "run()" operation (lines 03-05 in the code example below). A new thread is started by first creating a new *Thread* object named "thread2" (line 09) by passing an instance of the *MyThread* class to the *Thread* constructor on the right-handed side of line 09. When the thread object "thread2" is started by calling the "start()" operation (line 10), it begins executing by first calling the "run()" operation in lines 03-05.

```
01 | class MyThread extends Runnable
02 | { ...
03 | public void run()
04 | { ... // Include self-written code here
05 | }
06 | ...
07 | }
08 | ...
09 | Thread thread2 = new Thread(new MyThread());
10 | thread2.start();
11 | ...
```

There is no difference between the run-time behavior of both objects "thread1" and "thread2", whether created through a specialized class of *Thread* or the "Runnable" interface. But in the design phase, there is an important difference because Java does not allow multiple inheritance of classes so that a specialized class cannot inherit from another general class. If this is required, a thread class should be implemented using the "Runnable" interface.

What follows is an alphabetic list of *Thread* operations according to http://java.sun.com/products/jdk/1.1/docs/api/java.lang.Thread.html.

- **activeCount()**
  Returns the current number of active threads in this thread group.
- **checkAccess()**
  Determines if the currently running thread has permission to modify this thread.
- **currentThread()**
  Returns a reference to the currently executing thread.
- **destroy()**
  Destroys this thread without any cleanup.
- **getName()**
  Returns this thread's name.
- **getPriority()**
  Returns this thread's priority.
- **getThreadGroup()**
  Returns this thread's thread group.
- **interrupt()**
  Interrupts this thread.
- **interrupted()**
  Tests if the current thread has been interrupted.
- **isAlive()**
  Tests if this thread is alive.
- **join()**
  Wait for this thread to finish.
- **join(long)**
  Wait at most the specified milliseconds for this thread to finish.
- **resume()**
  Resume a suspended thread.
- **run()**
  If this thread was constructed using the Java "Runnable" interface, then the thread's "run()" operation is called; otherwise, this operation does nothing.
- **setPriority(int)**
  Changes the thread priority to a given value in the range of 1-10.
- **sleep(long)**
  Causes the currently executing thread to sleep (temporarily cease execution) for the specified number of milliseconds.

- **start()**
  Causes a thread to begin execution; the Java Virtual Machine calls the run method of this thread.
- **stop()**
  Forces the thread to stop executing.
- **suspend()**
  Suspends this thread.
- **yield()**
  Causes the currently executing thread to temporarily pause and allow other threads to execute.

10.2.4 Thread priorities and scheduling

Every Java thread has a **priority**. Threads with higher priority are executed in preference to threads with lower priority. When code running in some thread creates a new *Thread* object, the new thread inherits its priority from its creator thread. Class *java.lang.Thread* defines three constants denoting the minimum, maximum and default priority values of threads:

- public final static int MIN_PRIORITY = 1 ;
- public final static int MAX_PRIORITY = 10 ;
- public final static int NORM_PRIORITY = 5 ;

The default priority value of a thread is always five. To change this priority, a thread's "setPriority(int)" can be used that changes the thread priority to a specified value in the range of 1-10. The thread operation "getPriority()" returns a thread's priority.

Thread scheduling is the switching of execution from one thread to another according to a given plan. In general, Java thread scheduling is done in three different ways:
1. The currently executing thread is blocked or finishes.
2. A thread with a higher priority becomes available and begins executing before a thread with a lower priority.
3. A lower-priority thread is forced to suspend its execution in favor of a higher-priority thread. Such a re-scheduling event can take place, if the installed Java Virtual Machine (JVM) supports the concept of "time-slicing".

Not all JVMs support "time-slicing". Without "time-slicing", one thread in a group of equal-priority threads runs to completion without interruption by another thread in this group that gets no chance to start executing. The concept of "time-slicing" solves this problem by allocating a short processor time, a "time-slice" also called a "quantum", to each member of a thread group. At the end of a "quantum" the execution is taken away from a thread, even if it was not completed, and given to the next member of the thread group with equal priority.

The main Java thread operations to do scheduling are "join()", "yield()", "interrupt()", sleep()", "wait()", "suspend()" and "resume()", the latter two ones complementing each other (see the list in the last paragraph).

10.2.5 Thread synchronization

As introduced in paragraph 10.2.1 when describing the disadvantages of Java threads, monitors and locks ensure that only one thread at a time executes a critical code region so that thread synchronization can be realized. The **goal** of thread synchronization is to guarantee that once a thread starts executing an operation of an object, any other thread should be prevented from calling any operation of the same object that may cause a conflict for the first thread running the first operation. This is made possible by declaring all potentially conflicting class operations as "synchronized", which is a Java key word.

Every *java.lang.Thread* class object that owns a "synchronized" operation is also a monitor. The monitor does only allow one thread at a time to run a "synchronized" operation. This thread obtains the object's lock thus preventing other threads from doing the same or calling other "synchronized" operations of the same object. If there are several "synchronized" operations in one object, only one can be running at once. Other threads trying to invoke other "synchronized" operations have to wait. When a "synchronized" operation finishes executing, the object's lock is released and the monitor allows the waiting thread with the highest priority to call the same or another "synchronized" operation of the same object.

A thread inside of an object's "synchronized" operation may not be able to continue, thus it calls its own "wait()" operation. This helps to keep the processors and monitors free from idle threads. Such a thread waits in a queue with other threads to receive the object's lock permission to enter. After a thread completed a "synchronized" operation, the object passes this information to the waiting queue invoking its "notify()" or "notifyAll()" operations. This is a signal for the waiting threads to try to obtain the object's lock.

Monitors and locks do not require direct Java code to manipulate them. Everything is done through the "synchronized" key word. There are two methods of using the "synchronized" statement:

1. Declare a class operation as "synchronized", for example:
```
01 | public synchronized int add1(int a, int b)
02 | { ... }
```

2. Declare a compound statement (a code block) as "synchronized", for example:
```
01 | synchronized (add2)
02 | { ... }
```

where "add2" is the object reference to the compound statement. The code inside of the block in line 02 cannot be executed until the current thread obtains the lock of the object named "add2". These "synchronized" statements are similar to "synchronized" operations, but can be used in a much more detailed way because a statement can be much smaller in size than a whole operation.

When using "synchronized" operations or statements, one has to be careful that a Java program does not get into a **deadlock** situation where threads are blocking each other by trying to receive a lock another thread owns and vice versa. Such a deadlock situation happens in the following Java code example with "synchronized" operations declared in two classes *AClass* and *BClass*:

Table 10.1. Java deadlock program example in folder "Program07"

```
       ...
01 | void button1_actionPerformed(ActionEvent e) {
02 |      AClass acs = new AClass();
03 |      BClass bcs = new BClass();
04 |
05 |      acs.start();
06 |      bcs.start();
07 |
08 |      acs.ARun(bcs);
09 |      bcs.BRun(acs);
10 | }
11 |    ...
12 | class AClass extends Thread{
13 |      public synchronized void ARun(BClass bc){ bc.BFunction(); }
14 |    public void AFunction(){
15 |        System.out.println("AFunction started in AClass object!");
16 |        for ( int i=0; i<10; i++ ) { if (i==9) i = 0;
17 |            System.out.println("A Step "+i);
18 |            textField1.setText("A Step "+i); }
19 |    };
20 | }
21 |    ...
22 | class BClass extends Thread{
23 |      public synchronized void BRun(AClass ac){ ac.AFunction(); }
24 |    public void BFunction(){
25 |        System.out.println("BFunction started in BClass object!");
26 |        for ( int i=0; i<10; i++ ) { if (i==9) i = 0;
27 |            System.out.println("B Step "+i);
28 |            textField1.setText("B Step "+i); }
29 |    };
30 | }
       ...
```

In this Java deadlock program example, a *java.lang.Thread* class object named "acs" is created in line 02 and started in line 05 after performing an action (e.g. a mouse click) on the *java.awt.Button* object named "button1" (line 01). When calling operation "ARun(bcs)" in line 08, the thread object "acs" must first obtain its own lock before being able to begin running this operation, which holds as parameter a *BClass* object named "bcs" that is invoking its "BFunction()" operation (line 13). For this, "acs" needs also the lock of the thread object "bcs", which was created, started and calls as well operation "BRun(acs)" in the same way as "acs" did (lines 03, 06 and 09). Thus, the calls for "ARun(bcs)" and "BRun(acs)" are interleaved (lines 08-09) and both thread objects wait for the lock of the other thread to be released.

Deadlocks can be very hard to detect as there could he three or more threads involved and it may only occur when the thread scheduling happens to occur in a particular order and at a particular time relative to the calls for operations. In general, the only way to avoid deadlocks is through careful design and avoiding operation calls that might cause conflicts. Debugging programs where deadlock occurs is notoriously difficult, so it's best not to let the problem happen in the first place!

10.3 Two simple Java applets to compute number pi

In this paragraph, two simple Java applets without and with threads are described to compute number π numerically. They serve as a starting point to illustrate the techniques of decomposition and parallel computing with Java threads introduced before.

A common numerical integration formula to generate π is given as follows:

$$\tan(\pi/4) = 1 \ \Rightarrow \ \frac{\pi}{4} = \arctan(1) \Rightarrow \ \pi = \int_0^1 \frac{4}{1+x^2}\, dx \ . \tag{10.1}$$

A standard numerical method for evaluating this integral is to approximate it by a sum of rectangles with equal width, whereby the function value at the rectangle's midpoint is taken as the rectangle's height. Thus, number π can be numerically computed by the sum of rectangles in the area below the curve $4/(1+x^2)$ between x-values zero and one, which represents the decomposition of this problem. The more rectangles with a smaller width are chosen, the more accurate the calculation. A screenshot of the first Java applet with a sequential implementation of this numerical method without Java threads is given in Fig. 10.1.

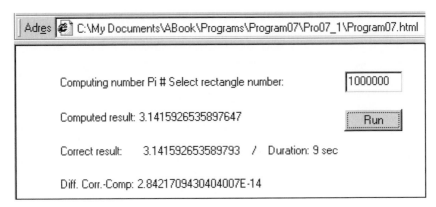

Fig. 10.1. Screen shot of the first applet to compute π without Java threads

Here the number of rectangles is set to 1 Million, which results in an error of about 2.8 E-14 with a calculation time of 9 sec on a 1 GHz PC with 128 MB RAM.

The Java code of this first applet to compute numerically number π is listed in the following table.

Table 10.2. Code in file "Program07.java" in folder "Program07/Pro07_1"

```
         ...
01  void button1_actionPerformed(ActionEvent e) {
02        label2.setText("Computing Started!  Please wait...");
03        long t0 = GetSeconds();
04        Double steps = new Double(textField1.getText());
05        long i, n = steps.longValue();
06        double sum = 0., w = 1./steps.doubleValue(), x;
07        for ( i=0; i<n; i++ ) {
08            x = w * (LongToDou(i) + 0.5);
09            sum = sum + Func(x); }
10
11            sum = sum/LongToDou(n);
12
13        label2.setText("Computed result: " + sum);
14        label3.setText("Correct result:      " +
15                    new Double(4.*Math.atan(1.)).toString() +
16                    "  /   Duration: " + (GetSeconds()-t0) + " sec");
17        label4.setText("Diff. Corr.-Comp: " + (4.*Math.atan(1.)-sum) );
18  }
19
20  double Func(double x) {
21        double y = 4./(1.+ x*x);
```

```
22 |       return y;
23 | }
24 |
25 | double LongToDou(long i) {
26 |       String s; s = "" + i;
27 |       Double d = new Double(s);
28 |       return d.doubleValue();
29 | }
30 |
31 | long GetSeconds(){
32 |       int Hour = Calendar.getInstance().getTime().getHours();
33 |       int Min  = Calendar.getInstance().getTime().getMinutes();
34 |       int Sec  = Calendar.getInstance().getTime().getSeconds();
35 |       long gs = Sec + 60*Min + 3600*Hour;
36 |     return gs;
37 | }
            ...
```

In this Java program example, a mouse click on the *java.awt.Button* object named "button1" (line 01) starts the code to compute π without threads. The number of rectangles inserted in the *java.awt.Textfield* object named "textfield1" is assigned to a *Double* object named "steps" in line 04. This value is also given to a variable "n" of type long (line 05) used inside of a for-loop to build the sum of rectangles in lines 07-09, whereby a rectangle's midpoint value "x" is computed in line 08. After completing this loop, the resulting summation variable "sum" of type double is divided by the number of steps "n" (line 11) and printed together with the computation time and the difference to the correct π value in lines 13-17.

The Java code of the second applet does the same with Java threads. It is listed in the following table followed by a screenshot in Fig. 10.2.

Table 10.3. Code in file "Program07.java" in folder "Program07/Pro07_2"

```
            ...
01 | void button1_actionPerformed(ActionEvent e) {
02 |       label2.setText("Computing Started!  Please wait...");
03 |       int i; long t0 = GetSeconds();
04 |       double sum = 0; Double steps = new Double(textField1.getText());
05 |
06 |       Compute t[] = new Compute[10];
07 |       for (i=0; i<=choice1.getSelectedIndex(); i++) {
08 |     t[i] = new Compute(i, choice1.getSelectedIndex()+1, steps.doubleValue());
09 |         t[i].start();
10 |         }
11 |
12 |       for (i=0; i<=choice1.getSelectedIndex(); i++) {
```

```
13      try { t[i].join(); sum = sum + t[i].sum;} catch (InterruptedException ex) {};
14      }
15          sum = sum/steps.doubleValue();
16
17      label2.setText("Computed result: " + sum);
18      label3.setText("Correct result:      " +
19              new Double(4.*Math.atan(1.)).toString() +
20              "  /   Duration: " + (GetSeconds()-t0) + " sec");
21      label4.setText("Diff. Corr.-Comp: " + (4.*Math.atan(1.)-sum) );
22  }
23
24  class Compute extends Thread {
25      public double sum; int id, maxThreads; double steps;
26    Compute(int id, int maxThreads, double steps) {
27          this.id = id; this.maxThreads = maxThreads; this.steps = steps; };
28
29      public void run() { sum = 0.;
30        long i, n = DouToLong(steps/IntToDou(maxThreads)), m =(id+1)*n;
31        double w = 1./steps, x; if (id==maxThreads-1) m = DouToLong(steps);
32        for ( i=id*n; i<m; i++ ) {
33            x = w * (LongToDou(i) + 0.5);
34            sum = sum + Func(x); }
35      }
36  }
37
38  double Func(double x) {
39      double y = 4./(1.+ x*x);
40      return y;
41  }
42
43  public double IntToDou(int i) {
44      String s; s = "" + i;
45      Double d = new Double(s);
46      return d.floatValue();
47  }
48  double LongToDou(long i) {
49      String s; s = "" + i;
50      Double d = new Double(s);
51      return d.doubleValue();
52  }
53  public long DouToLong(double d) {
54      String s; s = Double.toString(d);
55      Long i = new Long(s.substring(0,FindDot(s)));
56      return i.intValue();
57  }
58  public int FindDot(String s) {
```

```
59        int i, d = s.length();
60        for (i=0; i<s.length(); i++){
61            if (s.substring(i,i+1).equals(".")==true) d = i; }
62        return d;
63   }
64
65   long GetSeconds(){
66        int Hour = Calendar.getInstance().getTime().getHours();
67        int Min  = Calendar.getInstance().getTime().getMinutes();
68        int Sec  = Calendar.getInstance().getTime().getSeconds();
69        long gs = Sec + 60*Min + 3600*Hour;
70        return gs;
71   }
          ...
```

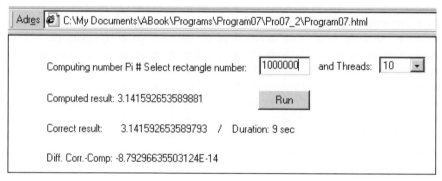

Fig. 10.2. Screen shot of the second applet to compute π with Java threads

Here the number of rectangles is set to 1 Million with 10 threads, which results in an error of about -8.8 E-14 with a calculation time of 9 sec on a 1 GHz PC with 128 MB RAM.

What are the **differences** between the code in Table 10.2 and 10.3 that uses Java threads to compute number π? The main differences can be found in lines 06-14 in Table 10.3 where objects of the Java thread class *Compute* are created, started and their results collected using the "join()" operation (line 13), as well as the code of class *Compute* in lines 24-36.

The Java thread class *Compute* has a generalization-specialization association with class *java.lang.Thread* and takes three parameters in its creator (lines 26-27 in Table 10.3): an ID named "id" starting with zero (line 07) as well as the number of threads called "maxThreads" and the number of steps called "steps". With these three parameters, a corresponding partial sum of the total sum of rectangles is computed in lines 29-35. For example, if "id" = 1, "maxThreads" = 2 and "steps"

= 10, then the for-loop in lines 32-34 runs with "n" = 10/2 = 5 and "m" = (1+1)*5 = 10 from five to ten. Note that "id" = 1 marks the second thread after "id" = 0.

Comparing the different computational times with and without threads, one can see in Fig. 10.1 and Fig. 10.2 that there is almost no time difference. The here used PC is a single-processor machine and multi-threaded programs run as fast as or even slower than conventional programs with sequential algorithms because of the thread creation, scheduling and synchronization, here realized through the "join()" operation, which causes a thread to wait until the current thread finishes its execution.

10.4 Creating a Java applet to compute satellite orbits

The objective of this paragraph is to create a Java applet for computing satellite orbits, before doing the same calculations in parallel with Java threads. In the beginning, the satellite orbit parameters and the satellite integration scheme are described in more detail. The starting point is **Kepler's laws**, which solve the two-body problem in classical mechanics (Heitz 1980), already mentioned in paragraph 1.4.2. In an inertial coordinate system S', the movement of two point masses q_1 and q_2 with only central symmetrical gravity fields and masses m_1 and m_2 can be described by the law of gravity as follows where k is the gravity constant and x'_i an unknown position vector in S' ($i \in \{1,2,3\}$ with x-, y-, z- coordinate values:

$$m_1 \, d^2 \, x'_{i\,1} / dt^2 = k \, m_1 \, m_2 \, x'_i / r^3 \, ,$$
$$m_2 \, d^2 \, x'_{i\,2} / dt^2 = - k \, m_1 \, m_2 \, x'_i / r^3 \, ,$$
$$x'_i = x'_{i\,2} - x'_{i\,1} \, , \quad r = \left| x'_i \right| = \sqrt{\delta_{ij} \, x'_i \, x'_j} \, , \qquad (10.2)$$

where δ_{ij} is the Kronecker delta tensor with ($i, j \in \{1,2,3\}$):
$\delta_{ij} = 1$ for $i = j$ and $\delta_{ij} = 0$ for $i \neq j$.

Division of the first equation in Equation (10.2) by m_1, the second by m_2 and building the difference between them results in:

$$d^2 \, x'_i / dt^2 = - k \, m \, x'_i / r^3 \, , \quad m = m_1 + m_2 \, . \qquad (10.3)$$

Formally, this equation describes the orbit of a point mass q_2 with $m_2 = 1$ around a central mass q_1 with a mass $m = m_1 + m_2$ in an inertial coordinate system with its origin in the mass center of q_1. The vector product of Equation (10.3) with the unknown position vector x'_i gives:

$$\varepsilon_{ijk}\, x_j'\, d^2\, x_k' / dt^2 = d\,[\,\varepsilon_{ijk}\, x_j'\, d\, x_k' / dt\,]/ dt \;=\; 0_i\;, \tag{10.4}$$

where the ε-tensor is defined by:
$\varepsilon_{ijk} = +\,1$ for i,j,k = 1,2,3 or 2,3,1 or 3,1,2;
$\varepsilon_{ijk} = -\,1$ for i,j,k = 1,3,2 or 2,1,3 or 3,2,1;
$\varepsilon_{ijk} = \quad 0$ for all other combinations of i,j,k.

Integration of Equation (10.4) gives Kepler's 2. Law:

$$\varepsilon_{ijk}\, x_j'\, v_k' = 2\, p_i = \text{constant}_i\;,\quad v_k' \;=\; d\, x_k' / dt\,. \tag{10.5}$$

The plane vector p_i is constant in time and always orthogonal to the position- and velocity vector. This also means that the orbital plane is constant in space and time. Twice of the length of the plane vector gives the plane constant:

$$\overline{p} \;=\; 2\,\big|\,p_i\big| \;=\; 2\; dA\,/\, dt \;=\; \text{constant}\,. \tag{10.6}$$

Kepler's 2. Law states that the position vector x'_i sweeps equal areas dA during equal times dt. The plane vector p_i holds three of the six unknowns of this problem. Two of them describe the orientation of the orbital plane. The best way to obtain the other three unknowns is to introduce polar coordinates (r,λ) in a second inertial coordinate system S, which has the z-axis in direction of p_i and the x-y-coordinate plane equal to the orbital plane so that position- and velocity vector can be expressed by:

$$x_i \;=\; r\,[\,\cos\lambda, \sin\lambda, 0\,]_i\;, \tag{10.7a}$$
$$v_i \;=\; r\,[\,\cos\lambda\;\; dr\,/\,dt - r\sin\lambda\;\; d\lambda\,/\,dt\,,$$
$$\quad\;\; \sin\lambda\;\; dr\,/\,dt + r\cos\lambda\;\; d\lambda\,/\,dt\,, \tag{10.7b}$$
$$\quad\;\; 0\,]_i\,,$$

With these identities Equations (10.3) and (10.6) change to:

$$d\,v_i\,/\, dt = -\,(\,k\, m\,/\, r^2\,)\,[\,\cos\lambda, \sin\lambda, 0\,]_i\;, \tag{10.8a}$$

$$\overline{p} \;=\; r^2\; d\lambda\,/\, dt \;\Rightarrow\; d\lambda\,/\, dt \;=\; \overline{p}\,/\, r^2\,. \tag{10.8b}$$

Inserting the second half of Equation (10.8b) into (10.8a) results in:

$$d\,v_i\,/\, d\lambda = -\,(\,k\, m\,/\,\overline{p}\,)\,[\,\cos\lambda, \sin\lambda, 0\,]_i\,. \tag{10.9a}$$

Integration over λ and using Equation (10.7b) gives for the x- and y-value of the velocity vector:

$$v_1 = \cos\lambda \ dr / dt - r \sin\lambda \ d\lambda / dt = -(k\,m / \overline{p}) \ \sin\lambda + A_1 , \qquad (10.9b)$$

$$v_2 = \sin\lambda \ dr / dt + r \cos\lambda \ d\lambda / dt = (k\,m / \overline{p}) \ \cos\lambda + A_2 . \qquad (10.9c)$$

A_1 and A_2 are integration constants. To eliminate dr / dt from these equations, one can build the difference (10.9b) $\sin\lambda$ - (10.9c) $\cos\lambda$ and insert the second half of Equation (10.8b), which results in **Kepler's 1. Law** as an analytical solution of Equation (10.3):

$$1/r = (k\,m / \overline{p}^{-2}) - (A_1 / \overline{p}) \cos\lambda - (A_2 / \overline{p}) \sin\lambda . \qquad (10.10)$$

The orbit of the point mass q_2 is a conic section with the central mass q_1 at one focal point of this conic section. The other three of the six total unknowns of this two-body problem are given by the geometry of the conic section, for example, the length of the semi-major axis a, the eccentricity e' or the mean anomaly M, describing the position of q_2 on the conic section at a time t_0. There are three types of conic sections possible:
1. Ellipse, if the total energy of q_2 is smaller than zero (e' < 1),
2. Parabola, if the total energy of q_2 is equal to zero (e' = 1), or
3. Hyperbola, if the total energy of q_2 is larger than zero (e' > 1),

whereby the total energy of q_2 is the sum of the kinetic and potential energy:

$$E_{Total} = E_{Kinetic} + E_{Potential} = (m / 2)\,v^2 - k\,m / r . \qquad (10.11)$$

Kepler's 3. Law is derived from the 1. and 2. one, stating that the ratio of the squared time period T^2 of one revolution of q_2 to the cubed semi-major axis a^3 is constant because π and (k m) are constant:

$$T^2 / a^3 = 4\pi^2 / (k\,m) = \text{constant} . \qquad (10.12)$$

Considering just ellipses (e' < 1), the previously introduced integration constants like p_i, A_1 or A_2 can be transformed into the **six Kepler orbit parameters** shown in Fig. 10.3 and listed below, which also represent the six unknowns of the two-body problem:

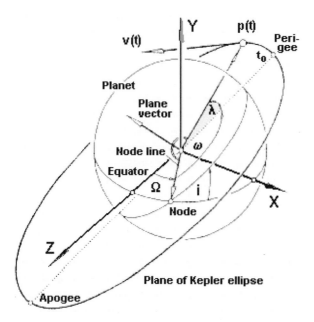

Fig. 10.3. The six Kepler orbit parameters: M, a, e', Ω, i and ω

1. M = Mean anomaly = $(k \, m \, / \, a^3 \,)^{1/2}$ $(t - t_0)$; M is proportional to λ, the angle between the perigee and the position x_i (t) of q_2 at time t according to Equations (10.7a-b); perigee is the closest point of q_2 to the central mass: x_i (t_0) = perigee; v_i (t) = velocity vector of q_2: v_i (t) = d x_i (t) / dt,
2. a = Semi-major axis of the ellipse,
3. e' = Eccentricity of the ellipse; e' = e/a; e = $(a^2 - b^2)^{1/2}$; b = semi-minor axis of the ellipse,
4. Ω = Right ascension of the ascending node = angle between the z-axis and the intersection between ellipse and equator called nodal line,
5. i = Inclination of the ellipse towards the equator = angle between the y-axis and the plane vector p_i, and
6. ω = Argument of perigee = angle between the nodal line and the direction towards the perigee.

Very common applications of Kepler's laws are to identify central and point mass with the sun and one of the nine planets in our solar system, or with the earth and an artificial earth satellite. The latter case will be further considered in the following applications.

The six Kepler orbit parameters provide an analytical solution to the two-body problem, which will be used to test a numerical integration scheme applied to the same problem, but in an earth-bound geocentric coordinate system. This scheme is based on the following formula of Runge-Kutta-Nyström 7-th order (Fehlberg 1969):

$$x_i^1 = x_i^0 + h\, v_i^0 + h^2 \sum_{\mu=0}^{12} c_\mu\, f_i^\mu\,,$$

$$v_i^1 = v_i^0 + h \sum_{\mu=0}^{12} d_\mu\, f_i^\mu\,,$$

$$f_i^0 \equiv f(t_0\,, x_i^0\,, v_i^0\,) \equiv a_i^0(t_0\,, x_i^0\,, v_i^0\,),$$

$$f_i^\mu = f(t_0 + \alpha_\mu\, h,$$

$$x_i^0 + \alpha_\mu h\, v_i^0 + h^2 \sum_{\nu=0}^{\mu-1} \gamma_{\mu\nu} f_i^\nu\,,$$

$$v_i^0 + h \sum_{\nu=0}^{\mu-1} \beta_{\mu\nu}\, f_i^\nu\,)\,,$$

(10.13)

where x_i^0 and v_i^0 are the position- and velocity vector of a point mass q_2 at a starting time t_0, h the step length, $t_1 = (t_0 + h)$ the next time step after t_0, and c_μ, d_μ, α_μ, $\beta_{\mu\nu}$, $\gamma_{\mu\nu}$ constant known coefficients. In this integration formula, the function values f_i^μ represent the acceleration a_i^μ and are computed at intermediate time steps $(t_0 + \alpha_\mu h)$ between t_0 and t_1 altogether thirteen different times ($\mu = 0, 1, 2, \ldots 12$).

The acceleration a_i^μ of a point mass q_2 in an earth-bound geocentric coordinate system for the two-body problem is given by:

$$a_i^\mu = -\,km_{Earth}\, x_i^\mu / \left| x_i^\mu \right|^3 - \varepsilon_{ijk} w_j\, \varepsilon_{klm} w_l x_m^\mu - 2\,\varepsilon_{ijk} w_j v_k^\mu\,,$$ (10.14)

where on the right-hand side, the
1. part is the acceleration through earth gravity, the
2. part the centrifugal acceleration, and the
3. part the coriolis acceleration.

Centrifugal and coriolis acceleration originate from the earth's rotation relative to the inertial coordinate system S described by the angular velocity w_i, where the earth's rotation axis is identical with the z-axis of the earth-bound geocentric coordinate system:

$$w_i = (2\,\pi/(24*3600))*(366.2422/365.2422)*[\,0,0,1\,]_i\,.$$ (10.15)

To **test** all equations introduced so far, another applet was created that integrates numerically hourly values of a position vector x_i and compares them with the correct analytical solutions gained from the six Kepler orbit parameters using the mean anomaly $M = (k\, m\, /\, a^3\,)^{1/2}\, (t - t_0\,)$, the only time-dependent orbit parameter, and an operation named "Kepler_to_Cartesian_Coord". In the beginning of the applet (line 08 in Table 10.4), this operation is also applied to compute the starting position- and velocity vectors x_i^0 and v_i^0 at time t_0 to be able to start the numerical integration scheme. The most important parts of the applet code are listed in the following table.

Table 10.4. Code in file "Program07.java" in folder "Program07/Pro07_3"

```
       ...
01 | public void init() { ...
02 |     textArea1.setText("Integrated 3D-Cartesian Satellite Coordinates"+"\n"+
03 |               "in an Geocentric Earth-Bound Coordinate System"+"\n\n"+
04 |               "The integrated XYZ-Coordinates and their Differences to the
05 |                   correct values"+"\n"+
06 |               "are in [m] at some given time points:");
07 |
08 |     Kepler_to_Cartesian_Coord( t, a, es, ra, in, ap, x, v);
09 |     for ( int i=0; i<ia; i++ ) { //******** Kepler Elements and Integration
10 |       Orbit_Integration(x,v,h,is); targ = targ + h*is;
11 |       sec=sec + DouToInt(h)*is; while(sec>=60){ min++; sec=sec-60; }
12 |     while(min>=60){ hour++; min=min-60; }
13 |       Kepler_to_Cartesian_Coord( t+targ, a, es, ra, in, ap, y, w);
14 |       for ( int j=0; j<=2; j++) { y[j] = y[j]-x[j]; }
15 |       textArea1.setText( textArea1.getText() +"\n\n"+
16 |                   year+" "+month+" "+day+" "+hour+": "+
17 |                   x[0]+" "+x[1]+" "+x[2]+"\n"+ "              "+
18 |                   y[0]+" "+y[1]+" "+y[2]);
19 |     }
20 | } // END of public void init() {
21 |
22 | //******** Transformation of Kepler Elements to Cartesian Coordinates
23 | void Kepler_to_Cartesian_Coord(double t, double a, double es, double ra,
24 |           double in, double ap, double x[], double v[]) { int i, j, k;
25 | double kn=Math.sqrt(kme/a/a/a), tu=2.*pi/kn, b=Math.sqrt(a*a*(1.-es*es)),
26 |     an, r, ve, vb, ce=2.*pi*a*b/tu, e = Math.sqrt(a*a-b*b), m,ae,ae0,ab,ab0;
27 |     double v1[] = new double [3]; double v2[] = new double [3];
28 |     double R[][]= new double [3][3];
29 |
30 |     m = kn*t; ae0 = m; ae = ae0 + es*Math.sin(ae0);
31 |     while (Math.sqrt((ae-ae0)*(ae-ae0))>1.e-14) {
32 |         ae0 = ae; ae = m + es*Math.sin(ae0); }
33 |     an = 2.*Math.atan2( Math.sqrt(1.+es)*Math.tan(ae/2.), Math.sqrt(1.-es) );
34 |       r = a*(1.-es*es)/(1.+es*Math.cos(an));
35 |       x[0]=r*Math.cos(an)+e; x[1]=r*Math.sin(an); x[2]=0.;
36 |       ab0=Math.atan2( x[1], x[0]);
37 |       ve = Math.sqrt( 1.+e/b*e/b*Math.cos(ab0)*Math.cos(ab0) );
38 |       ab =Math.atan2((x[1]+(a*a/b-b)/ve*Math.sin(ab0)), x[0]);
```

```
39
40         while (Math.sqrt((ab-ab0)*(ab-ab0)))>1.e-14) {
41             ve=Math.sqrt(1.+e/b*e/b*Math.cos(ab0)*Math.cos(ab0));
42           ab0 = ab; ab =Math.atan2((x[1]+(a*a/b-b)/ve*Math.sin(ab0)), x[0]); }
43        x[0]=r*Math.cos(an); vb = ce/( x[0]*Math.cos(ab) + x[1]*Math.sin(ab) );
44        v[0]= -vb*Math.sin(ab); v[1]= vb*Math.cos(ab); v[2]= 0.;
45
46         Mat_Rotat( ra/rho, in/rho, ap/rho,R);
47         Mat_Multi(v1,R,x); Mat_Multi(v2,R,v);
48         Mat_Rotat(0.,0.,GAST/rho - av[2]*targ,R);
49         Mat_Multi(x,R,v1); Mat_Multi(v,R,v2);
50
51       for (i=0; i<=2; i++) { j=i+1; k=i+2; if(j==3) j=j-3; if(k==3||k==4) k=k-3;
52          v2[i] = av[j]*x[k] - av[k]*x[j]; }
53       for (i=0; i<=2; i++) { v[i] = v[i] - v2[i]; }
54   }
55
56   //********* Orbit Integration with Runge-Kutta-Nystroem method 7th-order
57   void Orbit_Integration(double x[], double v[], double h, int is) {
58       int i, j, k, l=1; double t=0.;
59       double x0[] = new double [3]; double v0[] = new double [3];
60       double  a[] = new double [3]; double fx[] = new double [3];
61       double fv[] = new double [3]; double f[][] = new double [14][3];
62
63       for (i=0; i<=2; i++) { x0[i]=x[i]; v0[i]=v[i]; }
64       while ( 1 <= is ) {
65        Acceleration(x0,v0,a,t); for (i=0; i<=2; i++) { f[0][i]=a[i]; }
66        for (i=1; i<=13; i++) {
67         for (j=0; j<=2;  j++) { fx[j]=x0[j] + h*ca[i-1]*v0[j]; fv[j] = v0[j];
68         for (k=0; k<=i-1; k++) {
69           fx[j]=fx[j] + cb[i][k]*f[k][j]*h*h; fv[j]=fv[j] + cb[k][i]*f[k][j]*h;}}
70         Acceleration(fx,fv,a,t+ca[i-1]*h); for (j=0; j<=2; j++) {f[i][j]=a[j];} }
71        for (i=0; i<=2; i++) { x0[i]=x0[i] + h*v0[i]; for (j=0; j<=13; j++) {
72          x0[i]=x0[i] + cx[j]*f[j][i]*h*h; v0[i]=v0[i] + cv[j]*f[j][i]*h; } }
73        t = t + h; l++; }
74       for (i=0; i<=2; i++) { x[i]=x0[i]; v[i]=v0[i]; }
75         }
76
77   //************ Acceleration in an earth-bound coordinate system
78      void Acceleration(double x[], double v[], double a[], double t) {
79      int i, j, k; double r = Math.sqrt(x[0]*x[0] + x[1]*x[1] + x[2]*x[2]);
80
81      for (i=0; i<=2; i++) { a[i] = - kme * x[i]/r/r/r; }
82      for (i=0; i<=2; i++) { j=i+1; k=i+2; if(j==3) j=j-3; if(k==3||k==4) k=k-3;
83        a[i] = a[i] - 2. * ( av[j]*v[k] - av[k]*v[j] )
84        - ( av[j] * (av[i]*x[j]-av[j]*x[i])
```

85	-av[k] * (av[k]*x[i]-av[i]*x[k])); }
86	}
87	
88	//******* Last *--R.3(A3)*R.1(A2)*R.3(A1)--* first rotation matrix R(3,3)
89	// Rotation clockwise when watching from positive axis on to coordin. plane
90	void Mat_Rotat(double A3, double A2, double A1, double R[][]) {
91	double CA1, CA2, CA3, SA1, SA2, SA3;
92	CA1 = Math.cos(A1); CA2 = Math.cos(A2); CA3 = Math.cos(A3);
93	SA1 = Math.sin(A1); SA2 = Math.sin(A2); SA3 = Math.sin(A3);
94	R[0][0]= CA3*CA1-SA3*CA2*SA1;R[0][1] =-CA3*SA1-SA3*CA2*CA1;
95	R[1][0]=SA3*CA1+CA3*CA2*SA1;R[1][1]=-SA3*SA1+CA3*CA2*CA1;
96	R[0][2] = SA3*SA2; R[1][2] =-CA3*SA2;
97	R[2][0] = SA2*SA1; R[2][1] = SA2*CA1;
98	R[2][2] = CA2;
99	}
100	
101	//********** Vector Transformation
102	void Mat_Multi(double x[], double R[][], double y[]) { int i, j, k;
103	for (i=0; i<=2; i++) { x[i] = 0.; for (j=0; j<=2; j++) {
104	x[i] = x[i] + R[i][j] * y[j]; } }
105	}
	...

In Table 10.4, the numerical integration is calculated inside of a for-loop (lines 09-19), when the applet is initialized in lines 01-20. Before in line 08, the "Kepler_to_Cartesian_Coord" is called to compute the starting position- and velocity vectors x_i^0 and v_i^0 at time t_0 stored in arrays of type double named "x[]" and "v[]". This operation is defined in lines 23-54 and works as follows (Heitz 1980):

1. Compute iteratively the excentric anomaly $E = M - e'$ sin E in lines 30-32, where E is named "ae",
2. Compute λ in line 33, where λ is named "an",
3. Compute r in line 34,
4. Compute x_i and v_i in the elliptic plane in lines 35-44, and
5. Transform x_i and v_i into the earth-bound geocentric coordinate system in lines 46-53 with the help of two other operations "Mat_Rotat" to build a rotation matrix (lines 90-99), and "Mat_Multi" to multiply a vector with this rotation matrix (lines 102-105).

The integration for-loop in lines 09-19 in Table 10.4 calls an operation named "Orbit_Integration" (line 10), which is coded in lines 57-75 according to Equation (10.13), whereby "ia" is the length of the integration interval, "h" the step length and "is" the number of integration steps per integration interval so that "ia" = "h" * "is". This operation "Orbit_Integration" contains calls to an operation named "Acceleration" (lines 65+70), which is defined in lines 78-86 according to Equation (10.14). At each integration interval, the integrated hourly position vector x_i is compared with the correct analytical solution using again operation "Kepler_to_Cartesian_Coord" (lines 13). The difference vector named "y[]" is built in line 14 and printed in lines 15-18. A sample output of the applet is shown in Fig. 10.4, which indicates that the numerical integration works very well with errors below 1 cm.

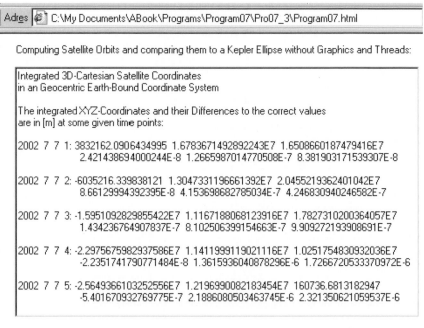

Fig. 10.4. Test of the numerical integration scheme with hourly differences between correct and integrated values below 1 cm

10.5 Object-oriented analysis and design

The multi-threaded Java applet for computing satellite orbits in parallel considered here should fulfill the following **requirements**. It should allow to:
1. Start Java threads to compute satellite orbits in parallel,
2. Stop any thread at any time as wanted,
3. Stop all threads at once if wanted, and
4. Plot the satellite orbits in parallel with different colors inside of the applet's graphics display.

As described later, the total number of threads will be restricted to a set maximum number (here it is 10), each number corresponding to a unique ID that represents a different satellite with a different orbit, each calculated and plotted in parallel. Such an applet executes one satellite integration program simultaneously by different threads, emulating a single instruction, multiple data (SIMD) machine.

A use-case diagram for these requirements is given in Fig. 10.5.

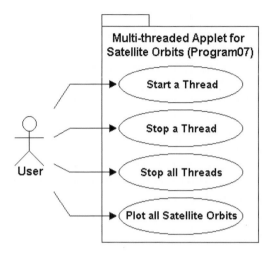

Fig. 10.5. Use-case diagram for a multi-threaded Java applet to compute satellite orbits

Apart from the Java applet in chapter 9, every application introduced so far was based on a 3-tier architecture consisting of user interface, processor and data manager, which were placed in different modules. Because applets are very small Java programs by nature and the application considered here is not very complex, the 3-tier approach was abandoned. Instead, all needed classes and objects are placed in one single module and Java file called "Program07.java", which has the same basic structure as "SimpleApplet.java" in Table 9.2. After importing all needed class libraries in the beginning, the applet class *Program07* is defined as an extension of class *java.applet.Applet* through a generalization-specialization association. Further, other classes are added to *Program07* through a whole-to-part association, which can be categorized as follows:

- Classes for graphical components in applet *Program07*, which are *java.awt.GraphClass*; *GraphClass* ∈ {*Button, Choice, Label*},
- Class *Satellite* as an extension of *java.lang.Thread* to compute and display satellite orbits.

All these added classes have a whole-to-part association with class *Program07* as shown in the class diagram below.

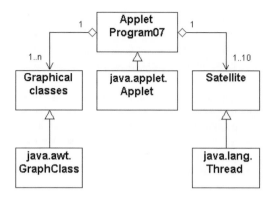

Fig. 10.6. Class diagram for the multi-threaded applet *Program07* to compute satellite orbits (n = 3)

The *java.awt* class package contains all needed classes for drawing graphical elements like lines inside of the applet's graphics display.

The scenario for the use-case diagram in Fig. 10.5 is shown in the following sequence diagram (Fig. 10.7).

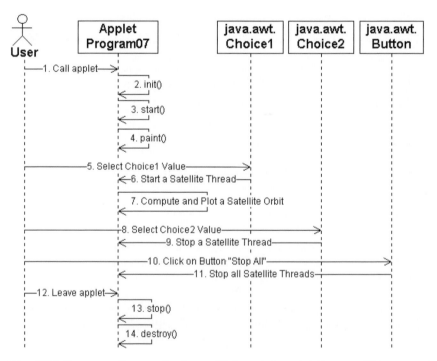

Fig. 10.7. Sequence diagram for the multi-threaded applet *Program07* to compute satellite orbits

In this diagram, a user calls an applet through an applet viewer or a Web browser in step 1. Each applet always begins with a series of three operation calls to "init()" in step 2, "start()" in step 3 and "paint()" in step 4. In the end when the user leaves the applet in step 12, the "stop()"and "destroy()" operations are called to finish the applet's lifecycle in steps 13-14. The "init()" operation is called once by the applet viewer or browser when an applet is loaded for execution. It initializes all applet objects. The "start()" operation is called after "init()" followed by the paint() operation, which may include the display of graphical components or the start of an animation.. The "stop()" operation is called when the applet should stop executing. The "destroy()" operation removes the applet from the computer memory – normally when a Web browser is closed.

10.6 Implementation of the designed model

The coded architecture of the multi-threaded applet to compute satellite orbits can be easily visualized with the JBuilder project explorer (Jensen et al. 1998) as shown in the figure below. This window is automatically displayed after opening the application's project file named "Program07.jpr" in folder "Program07/Pro07_4". All example programs can be **downloaded** at http://de.geocities.com/ bsttc2/book/SMOP.zip

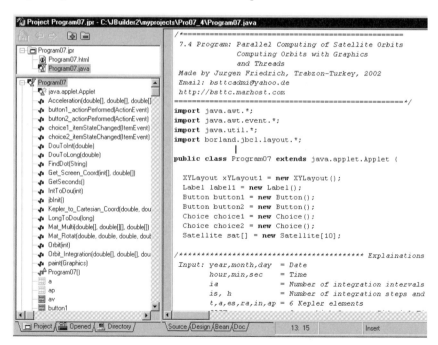

Fig. 10.8. JBuilder project explorer window for the multi-threaded applet *Program07* to compute satellite orbits with source view

As can be seen in the top left corner of this figure, the project "Program07" in file named "Program07.jpr" consists of just two files "Program07.html" and "Program07.java". In this implementation, the names used in the sequence diagram (Fig. 10.7) were retained as much as possible to make reading and understanding easier.

How "Program07" works in detail and how it was coded, can be more easily described by going through the application's sequence diagram (Fig. 10.7) and explaining step-by-step the corresponding program code. Before class *Program07* can be initialized and started in steps 2-4 of Fig. 10.7, all required classes inside of *Program07* have to be declared. They are visible in the source view on the right-handed side of Fig. 10.8 and listed in the following table.

Table 10.5. Class declarations in file "Program07.java" in folder "Program07/Pro07_4"

	...
01	XYLayout xYLayout1 = new XYLayout();
02	Label label1 = new Label();
03	Button button1 = new Button();
04	Button button2 = new Button();
05	Choice choice1 = new Choice();
06	Choice choice2 = new Choice();
07	Satellite sat[] = new Satellite[10];
	...

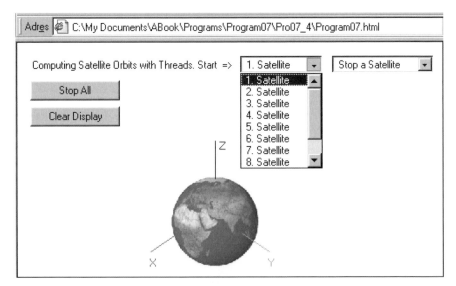

Fig. 10.9. Screen shot of the multi-threaded applet *Program07* to compute satellite orbits

In line 01 in Table 10.5, a *XYLayout* object "xYLayout1" is declared so that the *Program07* graphics display is based on an X-Y coordinate system as shown in Fig. 5.1. The *java.awt.Label* object "label1" (line02) is used to display the comment "Computing Satellite Orbits with Threads. Start =>" inside of the applet's graphics display as shown in Fig. 10.9. This figure indicates that selecting an item from the list of the *java.awt.Choice* object "choice1" (line 05) starts a satellite thread declared in line 07. A running thread can be stopped again by selecting it from the *java.awt.Choice* object "choice2" (line 06). Lines 03-04 declare two *java.awt.Button* objects named "button1, -2" labeled "Stop All" and "Clear Display" for stopping all running satellite threads and clearing the applet's graphics display.

The center of the applet's graphics display in Fig. 10.9 contains an image of the earth together with the earth-bound geocentric coordinate system introduced in paragraph 10.4. This is drawn through the "paint()" operation of class *Program07* after the applet was started (step 4 in the sequence diagram (Fig. 9.5)) as shown in the following table.

Table 10.6. "paint()" operation of class *Program07* in file "Program07.java" in folder "Program07/Pro07_4"

```
...
01  public void paint(Graphics g) {
02      Image world = getImage(getDocumentBase(), "World.gif");
03        g.drawImage(world, 250, 200, 100, 100, this);
04          g.setColor(Color.blue);
05          g.drawLine( 300, 202, 300, 160);
06          g.drawString( "Z", 305, 170); // Z-Axis in blue
07        g.setColor(Color.red);
08        g.drawLine( 256, 263, 226, 283);
09        g.drawString( "X", 225, 300);  // X-Axis in red
10          g.setColor(Color.green);
11          g.drawLine( 331, 263, 361, 283);
12          g.drawString( "Y", 360, 300); // Y-Axis in green
13  }
    ...
```

Line 01 shows that the "paint()" operation of class *Program07* holds one argument, the *java.awt.Graphics* object named "g", which represents the applet's graphics display. Only this object "g" and its drawing operations like "drawImage()", "drawLine()", "drawString()" and "setColor()" are needed to draw all graphical elements in Fig. 10.9 (lines 03-12).

The code for starting a satellite thread by changing an item of the *java.awt.Choice* object "choice1" (Fig. 10.9) according to steps 5-6 of the sequence diagram (Fig. 10.7) is listed in the table below.

Table 10.7. "Item State Changed" operation of the *java.awt.Choice* object "choice1" in file "Program07.java" in folder "Program07/Pro07_4"

```
      ...
01  void choice1_itemStateChanged(ItemEvent e) {
02       System.out.println(choice1.getSelectedIndex()+1 +". Sattelite started!");
03       int id = choice1.getSelectedIndex();
04       sat[id] = new Satellite(id);
05       sat[id].start();
06  }
      ...
```

The selected item index is assigned to an integer variable "id" in line 03, which is also the ID of a *Satellite* object "sat[]" created in line 04 and started in line 05. The code for class *Satellite* is given in the following table, which represents step 7 of the sequence diagram (Fig. 10.7).

Table 10.8. Class *Satellite* in file "Program07.java" in folder "Program07/Pro07_4"

```
      ...
01  class Satellite extends Thread {
02       int id;
03       Satellite(int id) {
04           this.id = id;
05       };
06         public void run() {
07           Orbit(id);
08           }
09  }
      ...
```

Class *Satellite* is derived from class *java.lang.Thread* by a generalization-specialization association in line 01. Its creator operation (lines 03-05) sets its "id" in line 04 to the value passed to the creator operation in line 04 in Table 10.7. Then the "run()" operation is defined in lines 06-08 with a call to the "Orbit" operation, which is listed in the table below.

Table 10.9. "Orbit" operation in file "Program07.java" in folder "Program07/Pro07_4"

```
      ...
01  void Orbit(int id) {
02       int s1[] = new int [3]; int s2[] = new int [3]; double rot = 0;
03       double x[] = new double [3]; double v[] = new double [3];
04       double y[] = new double [3]; double w[] = new double [3];
05       Graphics g = this.getGraphics();
```

```
06        if (id == 0) { g.setColor(Color.black);     rot = 0; }
07        if (id == 1) { g.setColor(Color.blue);      rot = 30; }
08        if (id == 2) { g.setColor(Color.cyan);      rot = 60; }
09        if (id == 3) { g.setColor(Color.magenta); rot = 90; }
10        if (id == 4) { g.setColor(Color.red);       rot = 120; }
11        if (id == 5) { g.setColor(Color.pink);      rot = 150; }
12        if (id == 6) { g.setColor(Color.orange);   rot = 180; }
13        if (id == 7) { g.setColor(Color.green);     rot = 210; }
14        if (id == 8) { g.setColor(Color.yellow);    rot = 270; }
15        if (id == 9) { g.setColor(Color.gray);      rot = 300; }
16
17        System.out.println("");
18        System.out.println("Integrated 3D-Cartesian Satellite Coordinates");
19        System.out.println("in an Geocentric Earth-Bound Coordinate System");
20        System.out.println("");
21   System.out.println("The integrated XYZ-Coordinates and their Differences
22                          to the correct values");
23        System.out.println("are in [m] at some given time points:");
24
25        hour = 0; t = 0; targ = t;
26        Kepler_to_Cartesian_Coord( t, a, es, ra+rot, in, ap, x, v);
27        System.out.println("");
28        System.out.println(year+" "+month+" "+day+" "+hour+": "+
29                  x[0]+" "+x[1]+" "+x[2]);
30        Get_Screen_Coord(s1, x);
31      for ( int i=0; i<ia; i++ ) {  //********** Kepler Elements and Integration
32        Orbit_Integration(x,v,h,is);  targ = targ + h*is;
33        sec=sec + DouToInt(h)*is; while(sec>=60){ min++; sec=sec-60; }
34        while(min>=60){ hour++; min=min-60; }
35        Kepler_to_Cartesian_Coord( t+targ, a, es, ra, in, ap, y, w);
36        Get_Screen_Coord(s2, x);
37        g.drawLine( s1[0], s1[1]+s1[2], s2[0], s2[1]+s2[2]);
38        for ( int j=0; j<=2; j++) { y[j] = y[j]-x[j]; s1[j] = s2[j]; }
39          System.out.println("");
40          System.out.println(year+" "+month+" "+day+" "+hour+": "+
41                  x[0]+" "+x[1]+" "+x[2]);
42          System.out.println("           "+
43                  y[0]+" "+y[1]+" "+y[2]);
44        if (i==ia-1) i = 0;
45      } // END of for ( int i=0; i<ia; i++ ) {
46 } // END of void Orbit(int id) {
47
48 //********** Get Screen Coordinates
49 void Get_Screen_Coord(int x[], double y[]) {
50        double Alpha = -30./rho, Gamma = 45./rho, sca = 100/12.8e+6;
51        double x0 = 300, y0 = 250;
```

```
52        double R[][]= {{Math.cos(Gamma),-Math.sin(Gamma)},
53              {Math.sin(Alpha)*Math.sin(Gamma),
54                Math.sin(Alpha)*Math.cos(Gamma)}};
55        x[0] = DouToInt(x0 - sca * (R[0][0]*y[0] +R[0][1]*y[1]) );
56        x[1] = DouToInt(y0 - sca * (R[1][0]*y[0] +R[1][1]*y[1]) );
57        x[2] = DouToInt(  - sca * Math.cos(Alpha)*y[2]);
58 }
   ...
```

The "Orbit" operation in Table 10.9 is very similar to the code in Table 10.4 with a few additional thread and graphical extensions for plotting satellite orbits. For this, the basis is the thread "id" passed to "Orbit" in line 01 and then used to assign different drawing colors and rotation variables of type double named "rot" in lines 06-15. The variable "rot" is then added to the right ascension of the ascending node "ra" in line 26 to compute different position- and velocity vectors x_i^0 and v_i^0 at the beginning of each thread so that different satellite orbits can be plotted in parallel in different colors.

The plotting of satellite orbits is done with two integer arrays "s1[]" and "s2[]" declared in line 02. The array "s1[]" is initialized in line 30 before the integration loop starts (line 30) using the "Get_Screen_Coord" operation (lines 49-58), which transforms 3D coordinates into screen coordinates with the same Equations (6.1) - (6.6) introduced in paragraph 6.1. At each step of the integration for-loop (lines 30-45), new screen coordinates "s2[]" are computed (line 36) and the differences plotted in the earth-bound geocentric coordinate system (line 37). Then "s2[]" is assigned to "s1[]" in line 38 to be able to draw the next part of the satellite orbit.

As in Table 10.4, the integration for-loop in lines 30-45 calls an operation named "Orbit_Integration" (line 32). Its code is listed in lines 57-54 in Table 10.4. At each integration interval, the integrated hourly position vector x_i is compared with the correct analytical solution using the operation "Kepler_to_Cartesian_Coord" (line 35). The code of this operation can be found in lines 23-54 in Table 10.4. The difference vector named "y[]" is built in line 38 and printed to the system console in lines 39-43. The integration for-loop is an unending loop because of line 44, which resets the for-loop variable "i" to zero as soon as it reaches the maximum value of "ia" minus one, whereby "ia" is the number of integration intervals.

A started *Satellite* thread object keeps running until stopped by the user according to steps 8-9 of the sequence diagram (Fig. 10.7). Selecting the corresponding "id" of the *Satellite* thread object from the list of the *java.awt.Choice* object "choice2" can do this using the following code.

Table 10.10. "Item State Changed" operation of the *java.awt.Choice* object "choice2" in file "Program07.java" in folder "Program07/Pro07_4"

```
    ...
01  void choice2_itemStateChanged(ItemEvent e) {
02      System.out.println(choice2.getSelectedIndex() +". Sattelite stopped!");
03      int id = choice2.getSelectedIndex()-1;
04      sat[id].stop();
05  }
    ...
```

The selected item index is assigned to an integer variable "id" in line 03, which is then used to call the "stop()" operation of the corresponding *Satellite* object "sat[]".

Instead of stopping one single *Satellite* thread object as done in Table 10.10, all running threads can be halted. The code for stopping all *Satellite* thread objects by clicking on the *java.awt.Button* object "button1" according to steps 10-11 of the sequence diagram (Fig. 10.7) is listed in the table below.

Table 10.11. "Action Performed" operation of the *java.awt.Button* object "button1" in file "Program07.java" in folder "Program07/Pro07_4"

```
    ...
01  void button1_actionPerformed(ActionEvent e) {
02      for ( int i=0; i<10; i++ )
03          if (sat[i].isAlive()==true) sat[i].stop();
04      System.out.println("All Sattelites stopped!");
05  }
    ...
```

In Table 10.11, a for-loop runs through all *Satellite* objects "sat[i]" from zero to ten in lines 02-03. If an object "sat[i]" is alive, it is stopped calling the thread's "stop()" operation.

Table 10.12. "Action Performed" operation of the *java.awt.Button* object "button2" in file "Program07.java" in folder "Program07/Pro07_4"

```
    ...
01  void button2_actionPerformed(ActionEvent e) {
02      Graphics g = this.getGraphics();
03      g.clearRect(0,0,800,600); paint(g);
04  }
    ...
```

If wanted the applet's graphics display can be cleared by a click on the *java.awt.Button* object named "button1". The corresponding code is listed in Table 10.12. This operation is helpful when the applet's graphics display gets filled up with plotted satellite orbits like shown in Fig. 10.10.

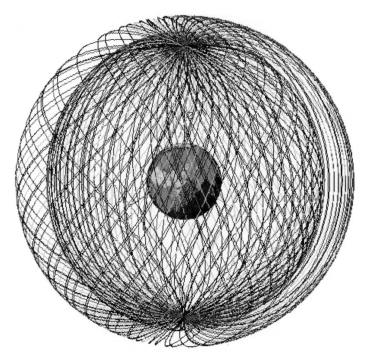

Fig. 10.10. Screen shot of a filled up *Program07* graphics display

10.7 Patterns used in this application

As already mentioned in paragraph 10.5, the 3-tier architecture used in all Visual Basic applications was abandoned here and replaced by a single module and Java file called "Program07.java". Nevertheless, the following **design patterns** are used because classes and objects in Java behave by their nature according to these patterns.

<u>**Creational Patterns:**</u>

2. Builder: Separate the construction of a complex object from its representation so that the same construction process can create different representations.

5. Singleton: Ensure a class only has one instance, and provide a global point of access to it.

Structural Patterns:

6. Adapter: Convert the interface of a class into another interface clients expect. Adapter lets classes work together that couldn't otherwise because of incompatible interfaces.

Behavioral Patterns:

14. Command: Encapsulate a request as an object, thereby letting you parameterize clients with different requests, queue or log requests, and support undoable operations.

19. Observer: Define a one-to-many dependency between objects so that when an object changes state, all its dependents are notified and updated automatically.

The **builder and command patterns** are applied in this applet in a simplified form when using the two *java.awt.Button* objects "button1" and "button2" labeled "Stop All" and "Clear Display" for stopping all running satellite threads and clearing the applet's graphics display. Instead of two different operations for a key press and a mouse click, which provide exactly the same functionality, only one reusable object or in a simpler way, just one single operation can be called. This simplified code as an alternative to Table 10.11 and Table 10.12 is shown in the table below.

Table 10.13. "Action Performed" and "Key Pressed" operations of the *java.awt.Button* objects "button1, -2" in file "Program07.java" in folder "Program07/Pro07_4"

```
      ...
01  void button1_actionPerformed(ActionEvent e) {
02        StopAll();
03  }
04  void button1_keyPressed(KeyEvent e) {
05        if (e.getKeyText(e.getKeyCode()).equals("Enter")==true)
06        StopAll();
07  }
08  void StopAll() {
09        for ( int i=0; i<10; i++ )
10            if (sat[i].isAlive()==true) sat[i].stop();
11        System.out.println("All Sattelites stopped!");
12  }
13
14  void button2_actionPerformed(ActionEvent e) {
15        ClearDisplay();
16  }
```

```
17 | void button2_keyPressed(KeyEvent e) {
18 |      if (e.getKeyText(e.getKeyCode()).equals("Enter")==true)
19 |      ClearDisplay();
20 | }
21 | void ClearDisplay() {
22 |      Graphics g = this.getGraphics();
23 |      g.clearRect(0,0,800,600); paint(g);
24 | }
       ...
```

The two operations simplifying the code in Table 10.13 are "StopAll" (lines 08-12) and "ClearDisplay" (lines 21-24). Both are called by the "Action Performed" and "Key Pressed" operations of the *java.awt.Button* objects "button1" and "button2" in lines 02+06 and lines 15+19, respectively.

The Singleton creational pattern is used in every applet because of class *java.applet.Applet*. This pattern ensures that only one object of the applet class, here *Program07*, exists in an applet.

The Adapter structural pattern and the Observer behavioral pattern are used, for example, when *java.awt.Button* objects respond to *ActionEvent* and Key*Event* objects using *ActionListener*, *KeyListener* and *KeyAdapter* objects as listed in the following table.

Table 10.14. Code for the *java.awt.Button* objects "button1, -2" in file "Program07.java" in folder "Program07/Pro07_4"

```
      ...
01 | button1.addActionListener(new java.awt.event.ActionListener() {
02 |      public void actionPerformed(ActionEvent e) {
03 |        button1_actionPerformed(e);
04 |      }
05 | });
06 | button1.addKeyListener(new java.awt.event.KeyAdapter() {
07 |      public void keyPressed(KeyEvent e) {
08 |        button1_keyPressed(e);
09 |      }
10 | });
11 |
12 | button2.addActionListener(new java.awt.event.ActionListener() {
13 |      public void actionPerformed(ActionEvent e) {
14 |        button2_actionPerformed(e);
15 |      }
16 | });
17 | button2.addKeyListener(new java.awt.event.KeyAdapter() {
18 |      public void keyPressed(KeyEvent e) {
```

19	button2_keyPressed(e);
20	}
21	});
	...

The *KeyListener* objects observe *KeyEvent* objects through the *KeyAdapter* objects (lines 06+17) and inform the *Button* objects permanently of any state changes of *KeyEvent* objects. The *KeyAdapter* object makes sure that *Button* objects can communicate with *KeyEvent* objects. The same is true for *ActionListener* and *ActionEvent* objects whereby *ActionListener* objects also take up the role of adapters (compare line 01 with line 06 and line 12 with line17).

10.8 Testing the application

The multi-threaded applet *Program07* to compute satellite orbits required much less testing than the previous applet *Program06*, mainly because of a missing database and much less graphical components. Further, large parts of the *Program07* code were reused and thus tested before in other applications, such as the code for:
1. Creating, starting, stopping and managing Java threads in Table 10.3 (file "Program07.java" in folder "Program07/Pro07_2"): a Java applet to compute number π numerically with threads,
2. Transforming the six Kepler orbit elements to Cartesian coordinates in operation "Kepler_to_Cartesian_Coord" and the code for the numerical integration scheme in Table 10.4 (file "Program07.java" in folder "Program07/Pro07_3"): a Java applet to compute satellite orbits without graphics and without threads, and
3. Transforming 3D Cartesian to screen coordinates in operation "Get_Screen_Coord" (Table 10.9) using the same Equations (6.1) - (6.6) introduced in paragraph 6.1 and listed in Table 6.6 (code for "PlotData" in file "MyFile.cls" in folder "Program03").

By reusing these code components, therefore, the time for implementing and testing the applet *Program07* could be immensely reduced.

Regarding the **performance** of *Program07*, there is normally no problem to compute and plot up to 10 satellite threads when directly executing "Program07.html"on a 1 GHz PC with 128 MB RAM used in this work to test all applications. But when RAM space gets less by opening too many programs at the same time, the response time is much more delayed when clicking on a *java.awt.Button* or *java.awt.Choice* object, for example. In such a situation, it can also easily happen that the *Program07* stops to respond to any user action.

Another factor affecting the performance of *Program07* is given in the "Orbit" operation in Table 10.9 where the difference vector named "y[]" is printed to the system console in lines 39-43. These print commands take up a lot of computation time. By deleting or deactivating them using the Java single-line comment sign "\\" as shown in the following Table 10.15, for example, the performance of the *Program07* applet can be substantially increased.

Table 10.15. "Orbit" operation with deactivated system print commands in file "Program07.java" in folder "Program07/Pro07_4"

	...
01	void Orbit(int id) {
...	...
38	for (int j=0; j<=2; j++) { y[j] = y[j]-x[j]; s1[j] = s2[j]; }
39	// System.out.println("");
40	// System.out.println(year+" "+month+" "+day+" "+hour+": "+
41	// x[0]+" "+x[1]+" "+x[2]);
42	// System.out.println(" "+
43	// y[0]+" "+y[1]+" "+y[2]);
44	if (i==ia-1) i = 0;
45	} // END of for (int i=0; i<ia; i++) {
46	} // END of void Orbit(int id) {
	...

Another important factor influencing the speed of *Program07* very much, is the length of the integration interval "ia" determined by the step length "h" multiplied by the number of integration steps per interval "is" (line 32 in Table 10.9 calling the "Orbit_Integration" operation in lines 57-75 in Table 10.4) so that "ia" ="h" * "is". The present settings in file "Program07.java" in folder "Program07/Pro07_4" are:

"h" = 30 sec, "is" = 2   =>   "ia" = "h" * "is" = 60 sec.

Both variables "h" and "is" can be easily used to manipulate the speed of the *Program07* applet. A larger value for "ia" accelerates the orbit computation and plotting, a smaller value slows it down, whereby the speed-up is more or less proportional to the value for "ia".

List of Abbreviations

ActiveX	Network version of OLE libraries
AWT	Java's Abstract Windowing Toolkit
CAL	Computer-Assisted Learning
CASE	Computer-Aided Software Engineering
CORBA	Common Object Request Broker Architecture
CRC	Class Responsibility Collaboration
DCOM	Distributed Component Object Model
DDL	Data Definition Language
DLL	Dynamic Link Library
DML	Data Manipulation Language
DTM	Digital Terrain Model
GIS	Geographical Information System
GRASP	General Responsibility Assignment Software Patterns
GUI	Graphical User Interface
JBCL	Java Beans Component Library
JDK	Java Developer's Kit
JVM	Java Virtual Machine
MFC	Microsoft Foundation Class
MIMD	Multiple Instruction, Multiple Data
OCL	Object Constraint Language
OID	Object Identifier
OLE	Object Linking and Embedding
OMT	Object Modeling Technique
OO	Object-Oriented
OOA	Object-Oriented Analysis
OOD	Object-Oriented Design
OOM	Object-Oriented Methodology
OOSE	Object-Oriented Software Engineering
OWL	Object Window Library
SIMD	Single Instruction, Multiple Data
SQL	Structured Query Language
UML	Unified Modeling Language
URL	Unified Resource Locator
VB	Visual Basic
WFM	Waterfall Model

List of Figures

List of Tables

Bibliography

Alexander, C., S. Ishikawa and M. Silverstein (1977) A Pattern Language -Towns-Building-Construction, Oxford University Press

Baujard, O., S. Pesty and C. Garbay (1994) MAPS: a language for multi-agent system design, Expert Systems Vol.11 No.2 pp.89-98

Boehm, B. (1988) A Spiral Model for Software Development and Enhancement, IEEE Computer May pp.61-72

Booch, G. (1994) Object-Oriented Analysis and Design with Applications, 2nd ed., Addison-Wesley Massachusetts

Booch, G. (1996) Object Solutions: managing the object-oriented project, Addison-Wesley Massachusetts

Booch, G., J. Rumbaugh and I. Jacobson (1998) Unified Modeling Language User Guide, Addison-Wesley Massachusetts

Borland (1994) Object Windows Programmer's Guide, Scotts Valley California

Brebbia, C.A. (1978) The Boundary Element Method for Engineers, Pentech Press London

Coad, P. and E. Yourdon (1991) Object-Oriented Analysis, 2nd. ed., Yourdon Press Computing Series, Prentice-Hall New Jersey

Coad, P. and E. Yourdon (1991) Object-Oriented Design, Yourdon Press Computing Series, Prentice-Hall New Jersey

Coad, P., D. North and M. Mayfield (1995) Object Models: Strategies, Patterns, and Applications, Prentice-Hall New Jersey

Coyle, F.P. (2000) Legacy Integration – Changing Perspectives, IEEE Software March/April pp.37-41

Cunningham, W. (2001) The History of Patterns, http://c2.com/cgi/wiki?HistoryOfPatterns

Deitel, H.M. and P.J. Deitel (1997) Java How to program, Prentice-Hall New Jersey

Deitel, H.M., P.J. Deitel and T. Nioto (1999) Visual Basic 6 How to program, Prentice-Hall New Jersey

Deitel, H.M., P.J. Deitel and S.E. Santry (2002) Advanced Java 2 platform: How to program, Prentice-Hall New Jersey

Falk,G. and W. Ruppel (1983) Mechanik Relativität Gravitation, Springer-Verlag Berlin

Fehlberg, E. (1969) Klassische Runge Kutta Formeln fünfter und siebenter Ordnung mit Schrittweiten-Kontrolle, Computing No.40 pp.93-106

Fenton, N. and S.L. Pfleeger (1996) Software Metrics: A Rigorous and Practical Approach, 2nd ed., Int. Thomson Press London

Fichman, R,G. and C.F. Kemerer (1992) Object-Oriented and Conventional Analysis and Design Methodologies, Computer Vol.25 No.10 pp.22-39

Friedrich, J. (1999) A Dual Reciprocity Boundary Element Model for the Degradation of strongly eroded Archaeological Signs, Mathematics and Computers in Simulation No.48 pp.281-293

Fritzon, D., P. Fritzon, L. Viklund, and J. Herber (1994) Object-oriented mathematical modelling applied to machine elements, Computers and Structures Vol.51 No.3 pp. 241-253

Gamma, E., R. Helm, R. Johnson and J. Vlissides (1995) Design Patterns, Elements of Reusable Object-Oriented Software, Addison-Wesley Massachusetts

Goldberg, A. and D. Robson (1984) Smalltalk 80 - The Language and its Implementation, Addison-Wesley Massachusetts

Grand, M. (1998) Patterns in Java: a catalog of reusable design patterns illustrated with UML, John Wiley & Sons Chichester UK

Hardwick, M. and D.L. Spooner (1989) The ROSE Data Manager: Using Object Technology to Support Interactive Engineering Applications, IEEE Trans. Knowledge and Data Engineering Vol.1 No.2 pp.285-289

Henderson-Sellers, B. (1997) OO Project Management: The Need for Process, IEEE Software July/August pp.96-97

Heiny, L. (1994) Windows Graphics Programming with Borland C++, John Wiley & Sons Chichester UK

Heitz, S. (1980) Mechanik fester Körper Bd.1, Dümmler Verlag Bonn

Hennessy, S., D. Twiggert, R. Driver, T. O'Shea, C.E. O'Malley, M. Byard, S. Draper, R. Hartley, R. Mohammed and E. Scanlon (1995) Design of a computer-augmented curriculum for mechanics, Int. Journal of Science Education Vol.17 No.1 pp.75-92

Horstmann, C. (1998) Computing Concepts with Java Essentials, John Wiley & Sons Chichester UK

Jacobson, I. (1992) Object-Oriented Software Engineering: A Use Case Driven Approach, Addison-Wesley Massachusetts

Jensen, C., B. Stone and L. Anderson (1998) JBuilder Essentials, McGraw-Hill New York

Kay, A. (1969) The reactive engine, PhD Thesis, University of Utah

Kecskemethy, A. and M. Hiller (1994) An object-oriented approach for an effective formulation of multi-body dynamics, Computer Methods in applied Mechanics and Engineering Vol.115 pp.287-314

Kerth, N., and W. Cunningham (1997) Using Patterns to Improve Our Architectual Vision, IEEE Software January/February pp.53-59

Koch,K.-R. (1988) Parameter Estimation and Hypothesis Testing in Linear Models, Springer-Verlag Berlin

Larman, C. (1998) Applying UML and Patterns: An Introduction to Object-Oriented Analysis and Design, Prentice-Hall New Jersey

McDermott, L.C. (1990) Research and computer-based instruction: Opportunity for interaction, American Journal of Physics Vol.58 No.5 pp.452-462

McDermott, L.C. (1991) Millikan Lecture 1990: What we teach and what is learned - Closing the gap, American Journal of Physics Vol.59 No.4 pp.301-315

McManus, J.P. (1999) Database Access with Visual Basic 6, SAMS Publisher Indianapolis Indiana

McMonnies, A. and W.S. McSporran (1995) Developing Object-Oriented Data Structures Using C++, McGraw-Hill New York

McMonnies, A. (2001) Visual Basic – an object oriented approach, Addison-Wesley Massachusetts

Monroe, R.T., A. Kompanek, R. Melton, and D. Garlan (1997) Architectual Styles, Design Patterns, and Objects, IEEE Software January/February pp.43-52

Nagin, P. and J. Impagliazzo (1995) Computer Science: a breadth-first approach with C, John Wiley & Sons Chichester UK

Norman, R.J. (1996) Object-Oriented System Analysis and Design, Prentice-Hall New Jersey

Popkin, Ltd. (2001) UML 1.1 Specification, UML Resource Center at www.popkin.com

Popper, K. (2002) The Logic of Scientific Discovery, Routledge London

Pressman, R.S. (1997) Software Engineering: A Practitioner's Approach, 4th ed., McGraw-Hill New York

Ragsdale, S (1992) Parallel Programming, McGraw-Hill Singapore

Rumbaugh, J., M. Blaha, W. Premerlani, F. Eddy and W. Lorensen (1991) Object-Oriented Modeling and Design, Prentice-Hall New Jersey

Sanal, Z. (1994) Finite Element Programming and C, Computers and Structures Vol.51 No.6 pp. 671-686

Schneider, D.I. (1998) An Introduction to Programming Using Visual Basic 5, 3rd ed., Prentice-Hall New Jersey

Stephens, R. (2000) Visual Basic Graphics Programming – Hands-On Applications and Advanced Color Development, John Wiley & Sons Chichester UK

Stroustrup, B. (1987) The C++ Programming Language, Addison-Wesley Massachusetts

Sully, P. (1993) Modelling the World with Objects, Prentice-Hall New Jersey

Visinsky, M.L., J.R. Cavallaro and I.D. Walker (1994): Expert system framework for fault detection and fault tolerance in robotics, Computers and Electrical Engineering Vol.20 No.5 pp.421-435

Winder, R. and G. Roberts (1998) Developing Java Software, John Wiley & Sons Chichester UK

Wirfs-Brock, R., B. Wilkerson and L. Wiener (1990) Designing Object-Oriented Software, Prentice-Hall New Jersey

Yourdon, E. and C. Argila (1996) Case Studies in Object-Oriented Analysis and Design, Yourdon Press Computing Series, Prentice-Hall New Jersey

Zachman, J. (1987) Framework for Information Systems Architecture, IBM Systems Journal Vol.26 No.3

Zienkiewicz, O.C. (1977) The Finite Element Method, 3rd ed., McGraw-Hill New York

Glossary

Abstract class
A class that can be used only as a superclass of some other class; no objects of an abstract class may be created except as instances of a subclass. An abstract class is typically used to define a common interface for a number of subclasses.

Abstract type
A type such that all objects conforming to it must also conform to one of its subtypes.

Abstraction
The act of concentrating the essential or general qualities of similar things; also, the resulting essential characteristics of a thing

Activity diagram
Activity diagrams are multi-purpose process flow diagrams that are used to model the behavior of a system. They can be used to model a use-case, or a class, or a complicated operation. An activity diagram is similar to a flow chart; the one key difference is that activity diagrams can show parallel processing. This is important when using activity diagrams to model system processes, some of which can be performed in parallel, and for modeling multiple threads in concurrent programs. Activity diagrams provide a graphical tool to model the process of use-cases. They can be used in addition to, or in place of, a textual description of the use-case, or a listing of the steps of the use-case. A textual description, code, or another activity diagram can detail the activity further.

Aggregation
A "whole-to-part" relationship between a superclass (or assembly class) representing the whole, and a subclass; a super class can have many aggregations at the same time. An aggregation is a special type of association and an alternative to other forms of generalization.

Analysis
An early phase of development, focused upon discovering the desired behavior of a system together with the roles and responsibilities of the central objects that carry out this behavior. It emphasizes questions of "what," rather than "how."

Applet
An applet is Java code that is automatically downloaded from the Web and runs within a Web browser on the user's desktop.

Application engineer
A person responsible for implementing the classes and mechanisms invented by the architect and abstractionists and who assembles these artifacts into small program fragments to fulfill the system's requirements.

Architect
A person responsible for evolving and maintaining the system's architecture; ultimately, the architect gives the system its conceptual integrity.

Architecture
A description of the organization and structure of a system; many different levels of architectures are involved in developing software systems, from physical hardware architecture to the logical architecture of an application framework.

Artifact
A tangible product of the development process

Association
A class of links with common structures; the general term for class connections

Behavioral prototype
An intermediate release of software used to explore alternative designs or to further analyze the dark corners of a system's functionality. Prototypes are distinctly not production-quality artifacts; hence, during the evolution of a system, their ideas but not their substance are folded into production architectural releases.

Class
A template for objects to encapsulate data, functionality, and behavior; a class consists of a set of objects that share a common structure and behavior. In the UML notation, a class is "a description of a set of objects that share the same attributes, operations, relationships and semantics".

Class diagram
As the objects are found, they can be grouped by type and classified in a class diagram. It is the class diagram that becomes the central analysis diagram of the object-oriented design, and one that shows the static structure of the system. The class diagram can be divided into different groups, e.g. for processing or data managing. It shows the classes involved with the user-interface, system logic, and data storage, for example.

Class member
An attribute (data) or operation (function, service, procedure,) within an object

Collaboration

The cooperation of a set of classes or objects, working together to carry out some higher-level behavior; the resulting behavior is not the responsibility of any one abstraction; rather, it derives from the distributed responsibilities of the community. Mechanisms represent collaborations of classes.

Collaboration diagram

Collaboration diagrams represent an alternative to sequence diagrams for modeling interactions between objects in the system. Whereas in the sequence diagram the focus is on the chronological sequence of the scenario being modeled, in the collaboration diagram the focus is on understanding all of the effects on a given object during a scenario. Objects are connected by links, each link representing an instance of an association between the classes involved. An association is also a class for defining a semantic relationship between classifiers, such as classes. The link shows messages sent between the objects, the type of message passed (synchronous, asynchronous, simple, balking, and time-out), and the visibility of objects.

Component

A discrete software module with an interface

Component diagram

Component diagrams allow to model and structure system domains, use-cases, classes, or components by using packages. In the UML, a package is the universal item to group elements, enabling builders to subdivide and categorize systems. Packages can be used on every level, from the highest level, where they are used to subdivide the system into domains, to the lowest level, where they are applied to group individual use-cases, classes, or components.

Concept

A category of ideas or things; A concept's intension is a description of its attributes, operations and semantics. A concept's extension is the set of instances or example objects that are members of the concept. Often defined as a synonym for type.

Conceptualization

The earliest phase of development focused upon providing a proof of concept for the system and characterized by a largely unrestrained activity directed toward the delivery of a prototype whose goals and schedules are clearly defined.

Conformance

A relation between types such that if type X conforms to type Y, then values of type X are also members of type Y and satisfy the definition of type Y.

Constraint

A restriction or condition on an element

Constructor
An operation to create objects

Container class
A class designed to hold and manipulate a collection of objects.

Contract
Defines the responsibilities and post-conditions that apply to the use of an operation. Also used to refer to the set of all conditions related to an interface.

Cost of ownership
The total cost of developing, maintaining, preserving, and operating a piece of software

Coupling
A dependency between elements (usually types, class, and subsystems), typically resulting from collaboration between the elements to provide a service

Delegation
The notion that an object can issue a message to another object in response to a message; the first object therefore delegates the responsibility to the second object

Deployment diagram
Deployment diagrams are used to model the configuration of run-time processing elements and the software components, processes, and objects that live on them. In the deployment diagram, you start by modeling the physical nodes and the communication associations that exist between them. For each node, you can indicate what component instances live or run on the node. You can also model the objects that are contained within the component. Deployment diagrams are used to model only components that exist as run-time entities; they are not used to model compile-time only or link-time only components.

Derivation
The process of defining a new class by reference to an existing class and then adding attributes and operations; the existing class is the superclass; the new class is referred to as the subclass or derived class.

Design
An intermediate phase of development, focused upon inventing architectures for the evolving implementation and specifying the common tactical policies that must be used by disparate elements of the system.

Destructor
An operation to eliminate objects

Discovery
The activity of investigation that leads to an understanding of a system's desired behavior and performance

Distributed system
A system whose graphical user interface (GUI), data, or computational elements are physically distributed

Distributed memory system
A system where each processor has its own private memory

Domain
A formal boundary that defines a particular subject or area of interest

Domain model
The sea of classes in a system that serve to capture the vocabulary of the problem space; also known as a conceptual model. A domain model can often be expressed in a set of class diagrams whose purpose is to visualize all of the central classes responsible for the essential behavior of the system, together with a specification of the distribution of roles and responsibilities among such classes.

Dynamic (or late) binding
The binding of virtual operations to an object when it is created during run-time; it is a concept to implement polymorphism by which the physical behavior of an object is prolonged to the adequate moment at run-time of the program.

Encapsulation
The combining of attributes (data) with the operations (functions) dedicated to manipulate the attributes. Encapsulation allows the hiding of information and is achieved by means of a new structuring and data-type mechanism, which is named "class". All interaction with an object is through a public interface of operations.

Essential minimal characteristics
The primary characteristics of a system, such as time to market, completeness, and quality against which economic decisions can be made during development.

Event
A noteworthy occurrence

Evolution
A later phase of development, focused upon growing and chancing the implementation through the successive refinement of the architecture, ultimately leading to deployment of the production system.

Extension
The set of objects to which a concept applies; the objects in the extension are the examples or instances of the concept

Framework
A set of collaborating abstract and concrete classes that may be used as a template to solve a related family of problems; it is usually extended via subclassing for application-specific behavior

Functional decomposition
The process of refining a problem solution by repeatedly decomposing a problem into smaller and smaller functional steps

Generalization
The activity of identifying commonality among concepts and defining supertype (general concept) and subtype (specialized concept) relationships; it is a way to construct classifications among concepts which are then illustrated in type hierarchies. Subtypes conform to supertypes in terms of intension and extension.

Idiom
An expression peculiar to a certain programming language or application culture, representing a generally accepted convention for use of the language. Idioms are a pattern that represents reuse in the small.

Implementation
The activity of programming, testing, and integration that leads to a deliverable application satisfying a system's desired behavior and performance.

Information management system
A system centered around a (usually distributed) domain model upon which user applications are built and which itself lives on top of a database layer that provides persistence, together with a communications layer that provides distribution.

Inheritance
A concept for expressing similarity by which a newly built, derived class inherits the attributes (data) and operations (functions) from one or more ancestor classes, while possibly redefining or adding new attributes and operations. This creates a hierarchy of classes like an inheritance tree.

Instantiation
The creation of an instance of a class

Invention
The activity of creation that leads to a system's architecture

Legacy system
An older, potentially moldy system that must be preserved for any number of economic or social reasons yet must also coexist with newly developed elements.

Link
A connection between two objects; an instance of an association in the same way that an object is an instance of a class

Mechanism
A structure that allows objects to collaborate in order to provide some behavior that satisfies a requirement of the problem; a mechanism is thus a design decision about how certain collections of objects cooperate. Mechanisms are a kind of pattern that represents the soul of a system's design.

Message
The mechanism by which objects communicate; usually a request to execute an operation

Methodology
A package of methods to perform a certain problem-solving strategy

Model
A description of static and/or dynamic characteristics of a subject area, portrayed through a number of views (usually diagrammatic or textual)

Multiplicity
The number of objects permitted to participate in an association

Object
Something you can do things to. An object has state, behavior, and identity; the structure and behavior of similar objects are defined in their common class. Thus, an object is also an instance of a particular class

Object diagram
Object diagrams encompass objects and their relationships at a point in time. An object diagram may be considered as a special case of a class diagram or a collaboration diagram.

Object-oriented analysis (OOA)
The investigation of a problem domain or system in terms of domain concepts, such as object types, associations, and state changes

Object-oriented design (OOD)
The specification of a logical software solution in terms of software objects, such as their classes, attributes, methods, and collaborations

Operation
In the UML, "a service that can requested from an object to effect behavior"; an operation has a signature, specified by its name and parameters; and it is invoked via a message. A method is an implementation of an operation with a specific algorithm.

Pattern
A common solution to a problem in a given context; a well-structured object-oriented architecture typically encompasses a range of patterns including idioms, mechanisms or frameworks.

Persistence
The enduring storage of the state of an object

Persistent object
An object that can survive the process or thread that created it; a persistent object exists until it is explicitly deleted.

Phase
Any of the major divisions of the development process (analysis, planning, design, implementation, and maintenance), encompassing a set of iterations each with a common economic objective

Polymorphism
A concept to change the behavior of operations when used by different classes; or in other words, an operation with a given name is shared up and down the class hierarchy, with each class in the hierarchy implementing the operation in a way appropriate to itself.

Post-condition
A constraint that must hold true after the completion of an operation

Pre-condition
A constraint that must hold true before an operation is requested

Private
A scoping mechanism used to restrict access to class members so that other objects cannot see them. Normally applied to all attributes, and to some methods.

Public
A scoping mechanism used to make members accessible to other objects. Normally applied to some methods, but not to attributes, since public attributes violate encapsulation.

Qualified association
An association whose membership is partitioned by the value of a qualifier

Query
A command that retrieves records from one or more database tables

Rational process
The development life cycle that leads from a statement of requirements to an implementation, achieved through a series of development phases whose purpose, products, activities, milestones, and quality measures are well defined and under steady control.

Real-time system
A system whose behaviors are tightly constrained by time and/or space

Receiver
The object that receives a message

Recursive association
An association where the source and the destination are the same object type

Responsibility
Some behavior for which an object is held accountable; a responsibility denotes the obligation of an object to provide a certain behavior and occasionally the delegation of that behavior to some other object.

Release
A stable, self-complete, and executable version of a system solution, together with any other peripheral elements necessary to use that release. Releases may be intermediate, meaning that they are meant only for internal consumption, or they may represent deployed versions, meaning that they are meant for external consumption.

Requirement
A user demand that a system should fulfill

Requirements specifications
A document describing what a software system does - its attributes and operations. Usually written from a user's point of view.

Role
The face that an object presents to the world; the same object may play different roles at different times and thus logically present a different face each time. Collectively, the roles of an object form the object's protocol.

Round-trip engineering
Or the round-trip model of system development, meaning the flexible process of continuous iteration back, forth and in between the different development phases analysis, planning, design, implementation and maintenance.

Scenario

An instance of a use case that signifies a single path through that use case; a scenario is defined as a time-ordered sequence of object interactions needed to fulfill a specific system responsibility.

Scenario diagram

A sequence diagram in pre-UML notations

Sender

The object that sends a message

Sequence diagram

Sequence diagrams are one of the most effective diagrams to model object interactions in a system. A sequence diagram is modeled for every use-case. Whereas the use-case diagram enables modeling of a processing view of a scenario, the sequence diagram contains implementation details of the scenario, including the objects and classes that are used to implement the scenario, and messages passed between the objects. Typically one examines the description of the use-case to determine what objects are necessary to implement the scenario. If you have modeled the description of the use-case as a sequence of steps, then you can "walk through" the steps to discover what objects are necessary for the steps to occur. A sequence diagram shows objects involved in the scenario by vertical dashed lines, and messages passed between the objects as horizontal vectors. The messages are drawn chronologically from the top of the diagram to the bottom; the horizontal spacing of objects is arbitrary.

Shared memory system

A system with a single pool of memory to which all processors have access

State

The condition of an object between events

State transition or shorter State diagram

While sequence and collaboration diagrams model dynamic actions between groups of objects in a system, the state diagram is used to model the dynamic behavior of a particular object, or class of objects. A state diagram is modeled for all classes deemed to have significant dynamic behavior. In it, you model the sequence of states that an object of the class goes through during its life in response to received stimuli, together with its own responses and actions. For example, an object's behavior is modeled in terms of what state it is in initially, and what state it transitions to when a particular event is received. You also model what actions an object performs while in a certain state. States represent the conditions of objects at certain points in time. Events represent incidents that cause objects to move from one state to another. Transition lines depict the movement from one state to another. Each transition line is labeled with the event that causes the transition. Actions occur when an object arrives in a state.

Strategic decision
A development decision that has sweeping architectural implications.

Strategy
A strategy consists of planned activities to reach pre-defined goals.

Subclass
A specialization of another class (the superclass); a subclass inherits the attributes and operations of the superclass.

Subtype
A specialization of another type (the supertype) that conforms to the intension and extension of the supertype

Superclass
A class from which another class inherits attributes and operations

Supertype
In a generalization-specialization relation, the more general type; an object that has subtypes

Synchronous message handling
The halting of further execution of instructions in a process until a message is sent or received

Thread
A sequence of code statements executed when a thread is started

Transition
A relationship between states that is traversed, if the specified event occurs and the guard condition met

Type
In the UML, descriptions of a set of similar objects with attributes and operations

Use case
A narrative, textual description of the sequence of events and actions that occur when a user participates in a dialog with a system during a meaningful process

Use-case diagram
A graphical representation of a use-case to capture the behavior of a system. Use-case diagrams consist of actors and use-cases. Actors represent users and other systems that interact with the system to be constructed. They are drawn as stick figures. They actually represent a type of user, not an instance of a user class. Use-cases represent the behavior of the system and what it goes through in response to

stimuli from an actor. They are drawn as ellipses and connected to actors through arrows.

Visibility
The ability to see or have reference to an object

Index

Printing: Mercedes-Druck, Berlin
Binding: Stein + Lehmann, Berlin